Reviews and critical articles covering the entire field of normal anatomy (cytology, histology, cyto- and histochemistry, electron microscopy, macroscopy, experimental morphology and embryology and comparative anatomy) are published in Advances in Anatomy, Embryology and Cell Biology. Papers dealing with anthropology and clinical morphology that aim to encourage cooperation between anatomy and related disciplines will also be accepted. Papers are normally commissioned. Original papers and communications may be submitted and will be considered for publication provided they meet the requirements of a review article and thus fit into the scope of "Advances". English language is preferred.
It is a fundamental condition that submitted manuscripts have not been and will not simultaneously be submitted or published elsewhere. With the acceptance of a manuscript for publication, the publisher acquires full and exclusive copyright for all languages and countries.
Twenty-five copies of each paper are supplied free of charge.

Manuscripts should be addressed to

Prof. Dr. F. **BECK**, Howard Florey Institute, University of Melbourne, Parkville, 3000 Melbourne, Victoria, Australia
e-mail: fb22@le.ac.uk

Prof. Dr. F. **CLASCÁ**, Department of Anatomy, Histology and Neurobiology,
Universidad Autónoma de Madrid, Ave. Arzobispo Morcillo s/n, 28029 Madrid, Spain
e-mail: francisco.clasca@uam.es

Prof. Dr. M. **FROTSCHER**, Institut für Anatomie und Zellbiologie, Abteilung für Neuroanatomie,
Albert-Ludwigs-Universität Freiburg, Albertstr. 17, 79001 Freiburg, Germany
e-mail: michael.frotscher@anat.uni-freiburg.de

Prof. Dr. D.E. **HAINES**, Ph.D., Department of Anatomy, The University of Mississippi Med. Ctr.,
2500 North State Street, Jackson, MS 39216–4505, USA
e-mail: dhaines@anatomy.umsmed.edu

Prof. Dr. N. **HIROKAWA**, Department of Cell Biology and Anatomy, University of Tokyo,
Hongo 7–3–1, 113-0033 Tokyo, Japan
e-mail: hirokawa@m.u-tokyo.ac.jp

Dr. Z. **KMIEC**, Department of Histology and Immunology, Medical University of Gdansk,
Debinki 1, 80-211 Gdansk, Poland
e-mail: zkmiec@amg.gda.pl

Prof. Dr. H.-W. **KORF**, Zentrum der Morphologie, Universität Frankfurt,
Theodor-Stern Kai 7, 60595 Frankfurt/Main, Germany
e-mail: korf@em.uni-frankfurt.de

Prof. Dr. E. **MARANI**, Department Biomedical Signal and Systems, University Twente,
P.O. Box 217, 7500 AE Enschede, The Netherlands
e-mail: e.marani@utwente.nl

Prof. Dr. R. **PUTZ**, Anatomische Anstalt der Universität München,
Lehrstuhl Anatomie I, Pettenkoferstr. 11, 80336 München, Germany
e-mail: reinhard.putz@med.uni-muenchen.de

Prof. Dr. Dr. h.c. Y. **SANO**, Department of Anatomy, Kyoto Prefectural University of Medicine,
Kawaramachi-Hirokoji, 602 Kyoto, Japan

Prof. Dr. Dr. h.c. T.H. **SCHIEBLER**, Anatomisches Institut der Universität,
Koellikerstraβe 6, 97070 Würzburg, Germany

Prof. Dr. J.-P. **TIMMERMANS**, Department of Veterinary Sciences, University of A
Groenenborgerlaan 171, 2020 Antwerpen, Belgium
e-mail: jean-pierre.timmermans@ua.ac.be

198
Advances in Anatomy, Embryology and Cell Biology

Editors

F. Beck, Melbourne · F. Clascá, Madrid
M. Frotscher, Freiburg · D.E. Haines, Jackson
N. Hirokawa, Tokyo · Z. Kmiec, Gdansk
H.-W. Korf, Frankfurt · E. Marani, Enschede
R. Putz, München · Y. Sano, Kyoto
T.H. Schiebler, Würzburg
J.-P. Timmermans, Antwerpen

Enrico Marani, Tjitske Heida,
Egbert A.J.F. Lakke,
and Kamen G. Usunoff

The Subthalamic Nucleus Part I: Development, Cytology, Topography and Connections

With 29 Figures

Enrico Marani
Tjitske Heida

Department of Biomedical Signals and Systems,
University of Twente,
7500 AE Enschede
The Netherlands

e-mail: e.marani@utwente.nl
e-mail: t.heida@el.utwente.nl

Egbert A.J.F. Lakke

Department of Neurosurgery,
Leiden University Medical Centre,
2300 RC Leiden
The Netherlands

Kamen G. Usunoff

Department of Anatomy & Histology,
Medical University Sofia,
1431 Sofia
Bulgaria

e-mail: uzunoff@medfac.acad.bg

ISSN 0301-5556
ISBN 978-3-540-79459-2 e-ISBN 978-3-540-79460-8

Library of Congress Control Number: 2008927199

© 2008 Springer-Verlag Berlin Heidelberg

This work is subject to copyright. All rights are reserved, whether the whole or part of the material is concerned, specifically the rights of translation, reprinting, reuse of illustrations, recitation, broadcasting reproduction on microfilm or in any other way, and storage in data banks. Duplication of this publication or parts thereof is permitted only under the provisions of the German Copyright Law of September 9, 1965, in its current version, and permission for use must always be obtained from Springer-Verlag. Violations are liable to prosecution under the German Copyright Law.

The use of general descriptive names, registered names, trademarks, etc. in this publication does not imply, even in the absence of a specific statement, that such names are exempt form the relevant protecttive laws and regulations and therefore free for general use.
Product liability: The publisher cannot guarantee the accuracy of any information about dosage and application contained in this book. In every individual case the user must check such information by consulting the relevant literature.

Printed on acid-free paper

9 8 7 6 5 4 3 2 1

springer.com

List of Contents

Abstract

This monograph (Part I of two volumes) on the subthalamic nucleus (STN) accentuates the gap between experimental animal and human information concerning subthalamic development, cytology, topography and connections. The light and electron microscopical cytology focuses on the open nucleus concept and the neuronal types present in the STN. The cytochemistry encompasses enzymes, NO, glial fibrillary acidic protein (GFAP), calcium binding proteins, and receptors (dopamine, cannabinoid, opioid, glutamate, γ-aminobutyric acid (GABA), serotonin, cholinergic, and calcium channels). The ontogeny of the subthalamic cell cord is also reviewed. The topography concerns the rat, cat, baboon and human STN. The descriptions of the connections are also given from a historical point of view. Recent tracer studies on the rat nigro-subthalamic connection revealed contralateral projections. Part II of the two volumes (volume 199) on the subthalamic nucleus (STN) starts with a systemic model of the basal ganglia to evaluate the position of the STN in the direct, indirect and hyperdirect pathways. A summary of in vitro studies is given, describing STN spontaneous activity as well as responses to depolarizing and hyperpolarizing inputs and high-frequency stimulation. STN bursting activity and the underlying ionic mechanisms are investigated. Deep brain stimulation used for symptomatic treatment of Parkinson's disease is discussed in terms of the elements that are influenced and its hypothesized mechanisms. This part of the monograph explores the pedunculopontine–subthalamic connections and summarizes attempts to mimic neurotransmitter actions of the pedunculopontine nucleus in cell cultures and high-frequency stimulation on cultured dissociated rat subthalamic neurons. STN cell models – single- and multi-compartment models and system-level models are discussed in relation to subthalamic function and dysfunction. Parts I and II are compared.

Abbreviations

A	Fields of Sano
A	Adenosine receptor
A8,A9	Catecholaminergic areas
ABC	Avidin-biotin-HRP complex
Alent	Ansa lenticularis (Fig. 2)
AMPA	α-Amino-3-hydroxy-5-methyl-4-isoxazole-proprionic acid
Am(g)	Amygdala
Apt	Anterior pretectal nucleus: AD, AM, AV indicate the dorsal, medial, ventral parts
APV	D-2-Amino-5-phosphono-valerate
AWSR	Array-wide spiking rate
AV	Anterior thalamic nucleus
BAPTA	1,2-bis(2-Aminophenoxy)-ethane-*N,N,N',N'*-tetraacetic acid
bc	Brachium conjunctivum
bci	Brachium of the colliculus inferior
BDA	Biotinylated dextran amine
BG	Basal ganglia
BI	Burst index
BIP	Burst intensity product
bp	Brachium pontis
CaBP	Calcium binding proteins
CB	Cannabinoid receptor
CB	Calbindin
CC	Corpus callosum
cd	Nucleus caudatus
Ce	Capsula interna (Fig. 2)
ChII	Chiasma opticum
CG	Central grey
Ci	Capsula interna (also Fig. 2)
ci	Capsula interna
Cl	Corpus Luysii

cl	Contralateral
cla	Claustrum
Cm	Corpus mamillare (Fig. 2)
CM	Centre median
Cml	Ganglion laterale corp. mamillare (Fig. 2)
Cmm	Ganglion mediale corp. mamillare (Fig. 2)
Coa	Commissural anterior (Fig. 2)
Coha	Commissura hypothalamica anterior (Fig. 2)
Cop	Commissura posterior
Cospm	Commissura supramamillaris (Fig. 2)
cp	Pedunculus cerebri
CR	Calretinin
Cu	Cuneiform nucleus
Csth	Corpus subthalamicum (Fig. 2)
ctb	Central tegmental tract of von Bechterew
ctt	Central tegmental tract
δ	Opioid receptor
d	Vesicle containing dendrites
D	Dopamine receptor
DA	Dopamine
Dbc	Decussation of brachium conjunctivum
DBS	Deep brain stimulation
dcv	Dense core vesicle terminals
DIV	Days in vitro
Dlx1/2	Homeobox gene
DNQX	6,7-Dinitroquinoxaline-2,3-dione
E	Embryonic day
EP	Nucleus entopeduncularis
F1	Flat type 1 (boutons)
F2	Flat type 2 (boutons)
Fhy	Fasciculus hypophyseos (Fig. 2)
Fmp	Fasciculus mamillaris princeps (Fig. 2)
Fo	Fornix (Fig. 2)
Fsp	Fasciculus subthalamico-peduncularis (Fig. 2)
fp	Fibrae perforantes (Fig. 2)
frtf	Fasciculus retroflexus Meynerti (Fig. 2)
Fu	Fasciculus uncinatus (Fig. 2)
GABA	γ-Aminobutyric acid
GAD	Glutamic acid decarboxylase
GAT	Specific high-affinity GABA uptake protein

GC	Gyrus cinguli
GCA	Gyrus centralis anterior
GCP	Gyrus centralis posterior
Gem	Ganglion ectomamillare (Fig. 2)
GF	Gyrus fusiformis
GH	Gyrus hippocampi
Ghb	Ganglion habenulae (Fig. 2)
gl	Corpus geniculatum
Glp	Glandula pinealis (Fig. 2)
glp	Globus pallidus (Fig. 2)
Glu	Ionotropic glutamate receptor
GP	Globus pallidus
GPe	Globus pallidus externus
GPi	Globus pallidus internus
H,h	H (Haubenfelder) fields of Forel (also Fig. 2)
5HT	5-Hydroxytryptamine
HRP	Horseradish peroxidase
HVA	High voltage activated currents
I	Insula Reilii (Fig. 2)
i	Nucleus internus gangl. med. corp. mamillaris (Fig. 2)
il	Ipsilateral
Ins	Insula
ISI	Interspike interval
κ	Opioid receptor κ
Kv3	Type delayed rectifier
L	Calcium channel type
ll	Lemniscus lateralis
Lm	Lemniscus medialis (also Fig. 2)
Lmi	Lamina medullaris interna
Lmm	Lamina medullaris medialis (Fig. 2)
Lml	Lamina medullaris lateralis (Fig. 2)
Lp	Posterior limitans thalamic nucleus
LPc	Gyrus paracentralis
LPi	Lobulus parietalis inferior
LR1	Large round type 1 (bouton)
LR2	Large round type 2 (bouton)
LTS	Low-threshold spike
μ	Opioid receptor μ
M,m	Cholinergic receptor

MEA	Midbrain extrapyramidal area
MEA	Multi-electrode array
mGlu	Metabotropic glutamate receptor
ml	Medial lemniscus
mlf	Fasciculus longitudinalis medialis
MPTP	1-Methyl-4-phenyl-1,2,3,6 tetrahydropyridine
mV	Motor nucleus of the nervus trigeminus
N	Calcium channel type
N	Substantia nigra
Nam	Nucleus amygdaliformis (Fig. 2)
Nans	Nucleus ansae lenticularis Meynerti (Fig. 2)
Narc	Nucleus arcuatus thalami (Fig. 2)
Nc	Nucleus caudatus (Fig. 2)
Nci	Nuclei of the colliculus inferior
NcM	Nucleus commissurae Meynerti (Fig. 2)
Ndd	Nuclei dorsales disseminati thalami (Fig. 2)
Neop	Nucleus of Darkschewitsch (Fig. 2)
NGF	Nerve growth factor
Ni	Substantia nigra
Nic	Substantia nigra pars compacta
Nir	Substantia nigra pars reticulata
Nkx-2.1	Homeobox gene
Nl	Nucleus centralis thalami (Fig. 2)
Nld	Nucleus lateralis dorsalis thalami (Fig. 2)
Nlv	Nucleus lateralis ventralis thalami (Fig. 2)
Nlve	Nucleus lateralis ventralis ext. thalami (Fig. 2)
Nlvi	Nucleus lateralis ventralis int thalami (Fig. 2)
Nm	Nucleus medialis thalami (Fig. 2)
Nmi	Nucleus mamilloinfundibularis (Fig. 2)
NMDA	*N*-Methyl-D-aspartate
NO	Nitric oxide
NOS	Nitric oxide synthase
NP	Pontine nuclei
Nso	Nucleus supraopticus
NR	Subtypes NMDA receptor
Ntg	Nucleus ruber tegmenti (Fig. 2)
Ntgd	Nucleus ruber tegmenti pars dorsalis (Fig. 2)
NIII	Nucleus oculomotorius
NVme	Mesencephalicus trigeminal nucleus
6-OH-DA	6 Hydroxy dopamine
ot	Tractus opticus
ω-CgTX	ω-Conotoxin
ω-AgTX	ω-Agatoxin

P	Postnatal day
P	Calcium channel type
pale	Globus pallidus externus
pali	Globus pallidus internus
parahip	Parahippocampal gyrus
PBP	Parabrachial pigmented nucleus
pc	Pedunculus cerebri
Ped	Pedunculus cerebri
Pl	Nucleus paralemniscalis
Pp	Pes pedunculus
PPN	Nucleus tegmenti pedunculopontinus
ppci	Capsula interna pars peduncularis
Pu	Putamen (Fig. 2)
Pul	Pulvinar (Fig. 2)
put	Putamen
PV	Parvalbumin
Q	Calcium channel type
R	Calcium channel type
R	Nucleus ruber
RE	Thalamo-reticular cells
RT	Nucleus reticularis thalami
Ru	Nucleus ruber
SC	Colliculus superior
SEM	Scanning electron microscopy
Sg	Suprageniculate nucleus
Shh	Sonic hedgehog
Smg	Gyrus supramarginalis
SN	Substantia nigra
SNc	Substantia nigra pars compacta
SNl	Substantia nigra pars lateralis
SNr	Substantia nigra pars compacta
Sns	Substantia nigra Soemmeringi (Fig. 2)
Spa	Substantia perforata anterior (Fig. 2)
SR	Small round boutons
St	Stria cornea (Fig. 2)
st	Spinothalamic tract
Stri	Stratum intermedium pedunculi (Fig. 2)
Strz	Stratum zonale thalami (Fig. 2)
STN	Subthalamic nucleus
T	Calcium channel type
t	Türck's part of cerebral peduncle

T1–3	Temporal gyri
TII	Tractus opticus
Tbc	Tuber cinereum (Fig. 2)
TC	Thalamo-cortical cells
TcTT	Tractus corticotegmentothalamicus Rinviki
TEA	Tetraethylammonium chloride
Tgpp	Nucleus tegmenti pedunculopontinus
Tpt	Tractus peduncularis transversus (Fig. 2)
Tri	Trigonum intercrurale
Tt	Taenia thalami
TTX	Tetrodotoxin
Un	Uncus
Va	Fasciculus mamillothalamicus (Fig. 2)
VA	Ventral anterior thalamic nucleus
VE	Nuclei ventralis thalami
Vim	Nucleus ventralis intermedius thalami
VM	Ventral medial thalamic nucleus
Voa	Nucleus ventro-oralis anterior
Vop	Nucleus ventro-oralis posterior
VPI	Nucleus ventralis posterior inferior thalami
VPL	Nucleus ventralis posterior lateralis thalami
VPM	Nucleus ventralis posterior medialis thalami
VTA	Ventral tegmental area
VIII	Ventriculus tertius (Fig. 2)
Wnt-3	Homeobox gene
Zi	Zona incerta (also Fig. 2)
II	Optic tract
3D	Three dimensional
IV	Nervus trochlearis

1 Introduction

1.1 Hemiballism

Hemiballism or hemichorea is a rare neurological disorder, but the crucial involvement of the subthalamic nucleus (STN) in its pathophysiology has been appreciated for decades (Jakob 1923; Martin 1927; Glees and Wall 1946; Whittier and Mettler 1949; Carpenter and Carpenter 1951; Crossman 1987). Only recently have serious doubts come forward. Postuma and Lang (2003) have described the STN as being involved in only a minority of cases, and indicated unrecognized causes such as non-ketotic hyperosmolar hyperglycaemia and complications of human immunodeficiency virus (HIV) infections. Moreover, the crucial involvement of a lesion of the STN is in doubt (Guridi and Obeso 2001; Postuma and Lang 2003). On the other hand, idiopathic Parkinson's disease (Battistin et al. 1996; Usunoff et al. 2002) is a common neurodegenerative disorder, but the key role of the STN in the pathophysiological origin of the parkinsonian state has become evident only recently (Miller and DeLong 1987; Mitchell et al. 1989; Bergman et al. 1990, 1994; Hollerman and Grace 1992; Guridi et al. 1993; Parent and Hazrati 1995b; Hassani et al. 1996; Levy et al. 1997, 2002; Blandini et al. 2000; Hirsch et al. 2000; Ni et al. 2000; Alvarez et al. 2001; Guridi and Obeso 2001; Magill et al. 2001; Marsden et al. 2001; Rodriguez-Oroz et al. 2001; Bevan et al. 2002; Houeto et al. 2002; Salin et al. 2002; Tintner and Jankovic 2002; Hamani et al. 2004). Surgery, primarily in the form of the bilateral, high-frequency stimulation of the STN (Benabid et al. 2000), is highly effective in parkinsonian patients who are responsive to levodopa but who experience marked motor fluctuation or other complications (Hamani et al. 2004; Tintner and Jankovic 2002; Perlmutter and Mink 2006; Kleiner-Fisman et al. 2006 and references therein). Houeto et al. (2002) point out that following STN stimulation, the parkinsonian motor disability improved by more than 60% (see also Sect. 2.3 of Part II of *The Subthalamic Nucleus*) and the levodopa equivalent daily dose was reduced by 60.5%. However, according to Houeto et al. (2002), the improvement in parkinsonian motor disability induced by STN stimulation is not necessarily accompanied by improvement of psychic function.

In a recent review paper (Temel et al. 2005), the involvement of the STN in the limbic and associative circuits is examined (Fig. 1). The authors report cognitive disorders such as altered verbal memory and fluency, altered executive functioning, changed attention behaviour, and disturbed working memory, mental speed and response inhibition after deep brain stimulation. The same holds for the limbic involvement of the STN. Changes in personality, depression, (hypo) mania, anxiety, and hallucinations are evident. The STN, therefore, not only possesses a key role in motor behaviour, but is also a "potent regulator" (Temel et al. 2005) in the limbic and associative circuits.

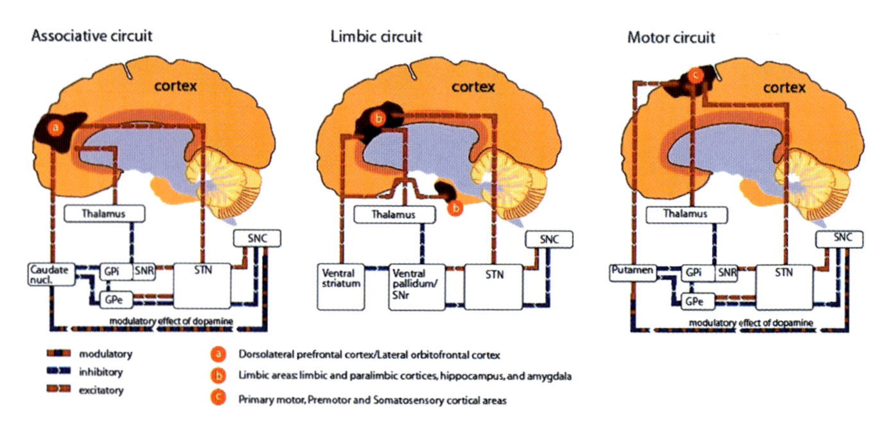

Fig. 1 The involvement of the STN in the associative, limbic and motor circuits according to Temel et al. (2005)

1.2 Early Subthalamic Research

The nucleus subthalamicus (Fig. 2), so named by Henle (1879) and also known as the corpus Luysii, was originally described by Luys (1865) as "bandelette accessoire de l'olive supérieure". The superior olive was the name for the red nucleus in Luys's descriptions.

> Luys saw the subthalamic nucleus as a centre for the dispersion of cerebellar influence upon the corpus striatum, a disposition that allows the nucleus to play a "crucial role in the synthesis of automatic motor actions". Hence, Luys not only discovered the subthalamic nucleus, but he was also the first to think about this structure as being intimately linked to the basal ganglia. Luys also traced the nervous fibres that link the subthalamic nucleus with the globus pallidus (the subthalamopallidal connection of the current literature) and described a fibre projection from the cerebral cortex to the subthalamic nucleus. He also clearly envisaged that the various areas of the cerebral cortex are directly represented at the level of the striatum via the corticostriatal projections (les projections corticostriées). (Parent 2002)

Several other names have been used for this area: nucleus of Forel by Edinger, lentiform disc by Meynert and the nucleus amygdaliformis by Stilling (see Dejerine 1901) or nucleus hypothalamicus (Villiger 1946). From 1865 until 1940 mainly the names corpus Luysii and/or nucleus subthalamicus were used (see e.g. Winkler 1928) and even nowadays corpus Luysii is still found in descriptive papers.

It was Forel (1872) who appreciated the nucleus, since his description of the fibre and connection fields (h, h1, h2) brought the nucleus to his attention; he gave it an anatomical description.

The nucleus in man ranges from 6–7.5 × 10 mm to 13 × 3–4 mm (Dejerine 1901). This long triangular or spindle-shaped nucleus is surrounded by fibre systems; its base is on the internal capsule with passing fibres of the "Kammsystem" (comb system) of Edinger, while on the other side fibre connections pass and join at its lateral pole. This other side of the STN is covered by the fasciculus lenticularis (h2 field of Forel, see Fig. 2). In his connective studies of the nucleus ruber of rabbit, cat, dog and

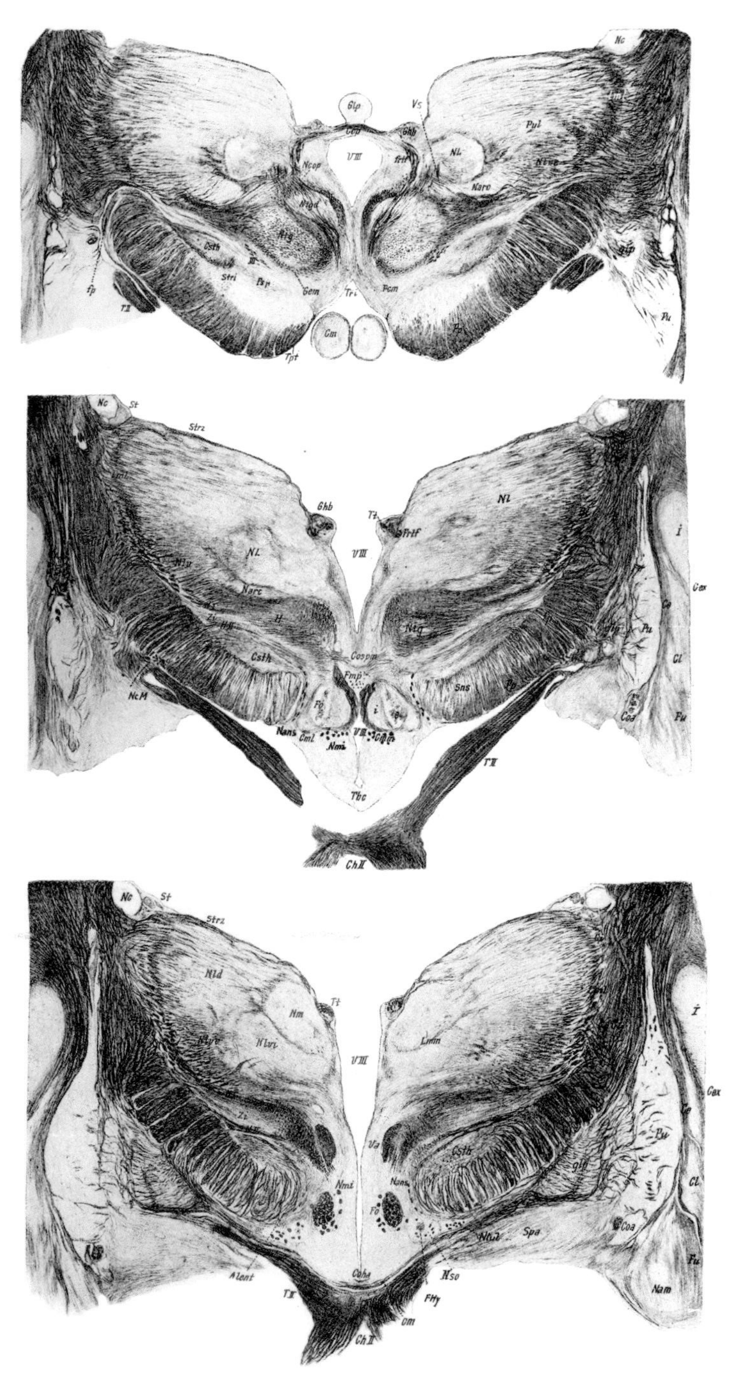

Fig. 2 Three transversal sections through the subthalamic area taken from Marburg (1910). *Csth*: corpus subthalamicus; *H*, *H1*, *H2*, fields of Forel; *Pp*, pedunculus cerebri; *Zi*, zona incerta; *fp*, fibrae perforantes (comb system of Edinger). For the other abbreviations see abbreviations list. Upper figure is 1.3× the other two figures. NB: The abbreviations are strictly according to the original terminology as used by Marburg

man, Von Monakov (1909) made or had lesions that included the subthalamic region. His main comment is that lesions in the sub- and hypothalamic area do not bring up long degenerating systems, and only very few short degenerating fibres were found that descend into the mesencephalic areas and not lower. In fact, Von Monakov (1909) restricts the subthalamic connections to the mesencephalic/diencephalic regions, denying cortical connections (see also Kappers et al. 1936), and he speculated on a pallidal connection.

Neuronal studies of the STN brought forward that two types of cells are predominantly present: small and large neurons. The large neurons should contain the same pigment as those from the substantia nigra (Winkler 1928). Interestingly, stimulation of frontal and medial parts of the nucleus gave dilatation of the pupil (Karplus and Kreidl 1909). In 1949, however, Hess described the area in which pupil dilatation is possible by stimulation. Pupil motoric symptoms are found already in tectum and mesencephalic grey. Pupil dilatation is found in his study in the STN and the anterior hypothalamus. Its anterior border is sharp at the area preoptica. The tegmentum gives no pupil dilatation. Therefore, the posterior border for pupil dilatation is at the transition mesencephalon–diencephalon. A parasagittal plane lateral of the red nucleus asks for a rather high stimulus. This area is therefore no longer considered as contributing to pupil dilatation. Pupil contraction is found in the middle of the thalamus. Hess (1949) therefore, concludes that the pupil dilation area is not restricted to the STN, while the pupil contraction is localized clearly more dorsally.

1.3 Ballism and the Subthalamic Nucleus

"Ballism(us) is characterized by the fairly abrupt onset of more or less extensive, vigorous, rapidly executed, poorly patterned, non-adaptive and seemingly purposeless activities of appendicular, truncal and/or faciocephalic striated muscles" (Meyers 1968). The term "hemiballismus" was introduced by Kussmaul according to Meyers (1968). Ballism in Greek means to throw and Kussmaul was impressed "by the gross resemblance of the involuntary limb movements to normal throwing activities" (Meyers 1968).

The first descriptions of (hemi)ballism were given by Greiff (1883) and Greidenberg (1882). Greiff's paper was on hemichorea and concerned ballism. Cortical bleedings and their consequent lesions (encephalomalacia; less frequently, intracerebral haemorrhages) were held responsible. Greidenberg (1882) noticed for the first time that the contralateral subthalamic *area* was involved. The involvement of a lesion of the subthalamic *nucleus* was for the first time noticed by Touche (1901).

An ipsilateral lesion of the STN in ballism was favoured by Fischer (1911), but Jakob (1923) found the lesions contralaterally in several cases. In various studies concerning 56 cases the primary lesions were found in 40 cases in the contralateral STN. In the other 16 cases secondary changes were noticed in contralateral afferent and efferent tracts of the STN. In most cases it concerned lesions of putamen or pallidum. Thus, at least in 71% of the cases that were investigated histologically, primary lesions of the STN were detected (Hallervorden 1957). Moreover, hemiballistic involuntary movements, also known as "choreoathetoid movements vaguely reminiscent of ballism" (Meyers 1968), can be brought about in rhesus monkeys after circumscribed lesion of the STN (Whittier 1952; Whittier and Mettler 1949; Carpenter et al. 1950). However, the modern cases studied with computed tomography (CT) or magnetic resonance imaging (MRI) showed from a total of 120 cases that 20% had no lesion at all and 53% had lesions outside the STN. Only 26% of the patients had a lesion in (18%) or possibly in (8%) the STN. Postuma and Lang (2003) concluded that "early reports may have been biased towards finding lesions in the STN".

Heredodegenerative cases were described later. The lesion bilaterally involves the STN. One case by Rakonitz (1933) and one case by Titica and van Bogaert (1946) showed pallido-subthalamic atrophy for such cases. Before 1987, 13 such cases had been reported (Hoogstraten et al. 1986). Pallido-subthalamic atrophy is seldom found in cases with torticollis and spastic trembling of the head.

A series of hemiballistic cases has been reported in which the STN was not damaged. The lesioned areas described were: afferent and efferent fibres of the STN, corpus striatum, thalamus, postcentral gyrus and multiple lesions (for an overview of these cases: see Meyers 1968). "Ballism appears to result from disruption of an extensive neuronal assembly or network, the topological details of which remain as yet vaguely identified. Such disruption evidently may occur at any of a number of critical nodes of the network" (Meyers 1968). On the other hand lesions of the STN have been described in parkinsonian patients without hemichorea or ballism (Guridi and Obeso 2001). It is nowadays, due to the extensive research, generally accepted that the STN is one of the most critical if not *the* critical node in the "ballism" network. This node in the "ballism" network nowadays receives increasing attention, since the STN is involved in deep brain stimulation (DBS) in Parkinson's disease and movement disorders (Hamani et al. 2004). The idea was to interrupt the excitatory influence of the STN to overcome the parkinsonian symptoms. STN lesions produce the unwanted ballism (but see criticism of Guridi and Obeso 2001), therefore researchers used a "reversible" lesion and placed electrodes for DBS into the STN (see Surmeier and Bevan 2003). The pioneering work of Benabid et al. (1989, 2000; Benabid 2003) and Benazzouz et al. (2000a, b, 2002) demonstrated that high frequency stimulation of the STN indeed resulted in improvement of resting tremor, rigidity and bradykinesia in Parkinson's disease and other movement disorders (Gross and Lozano 2000; Obeso et al. 2001).

This monograph will concentrate on the structure, development and connections of the STN and related mesencephalic nuclei as critical for the mathematical models that have been underestimating and only partially incorporating these connections in the majority of cases. These models are used for understanding high-frequency stimulation. Therefore, vol. 199 in this series will direct itself to the type of models used for simulation of subthalamic function and will give recent modelling results from our Biomedical Signals and Systems department.

2 Cytology of the Subthalamic Nucleus

In the discussion on the cytology of the STN two main themes are present: first, is the nucleus closed or open? This means, are the dendritic branches restricted to the nuclear area (closed) or do the dendrites reach out to other areas (open) (Mannen 1960)? The second point of discussion is the presence of interneurons within the STN.

The original studies on the cytology of the STN were carried out by Forel (1877; see Fig. 3). Ramon y Cajal (facsimile from 1955) demonstrated with the Golgi technique that the neurons in the STN are multipolar with pigment, spindle shaped or polygonal. The neurons bring forward long dendrites with spines that branch

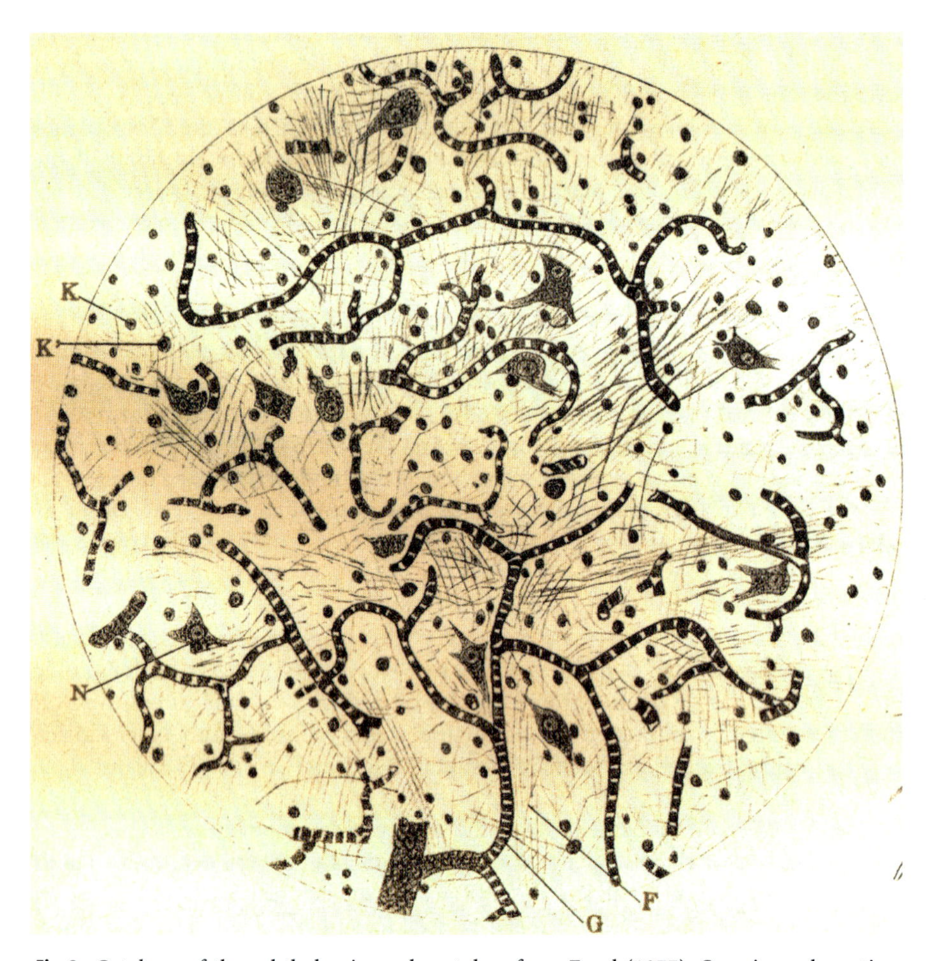

Fig. 3 Cytology of the subthalamic nucleus taken from Forel (1877). Carmine colouration; *N*, neurons; *G*, blood vessels; *F*, fine nerve fibres without organization; *K*, small granula; *K'*, large granula. "The dendrites are difficult to stain, as is normal in this region," noted Forel (1877) in the figure text

regularly. The initial axonal segment bows regularly, by which the axons are difficult to follow, towards bundles of descending fibres. An analogous description can also be found in Dejerine's *Anatomie des centres nerveux* (1901), in which he explicitly indicates that most of the neurons are of the Golgi type I as described by Kölliker (1891, 1896). In Winkler (1928) at least two types of human subthalamic neurons are discerned: parvo and magnocellular (Fig. 4). The magnocellular spindle-shaped neurons are on average two to three times larger in perikaryon diameter. Moreover, Winkler (1928) noticed that the magnocellular neurons in the medial part of the nucleus are smaller than those in its lateral part. The STN in its ventro-medial part connects to the substantia nigra, and since the subthalamic magnocellular neurons contain pigment, it is difficult to border both nuclei. The small neurons are placed in a "gelatine-like substance rich of myelinated fibres".

The pigment in the subthalamic neurons belongs to lipofuscin granula in *Macaca mulatta* and *M. nemestrina (*Rafols and Fox (1976) and seemingly not to neuromelanin as Winkler (1928) indicated (see also Usunoff et al. 2002). Lipofuscin granula show a dispersed presence in *M. mulatta* and an overall presence in *M. nemestrina* (for lipofuscin information, see Marani et al. 2006).

The question of an open or closed nucleus has been studied in several articles. Going from rat to man, the nucleus evolves from an open to a closed nucleus (see below). In rat the nucleus is considered clearly an open nucleus, since

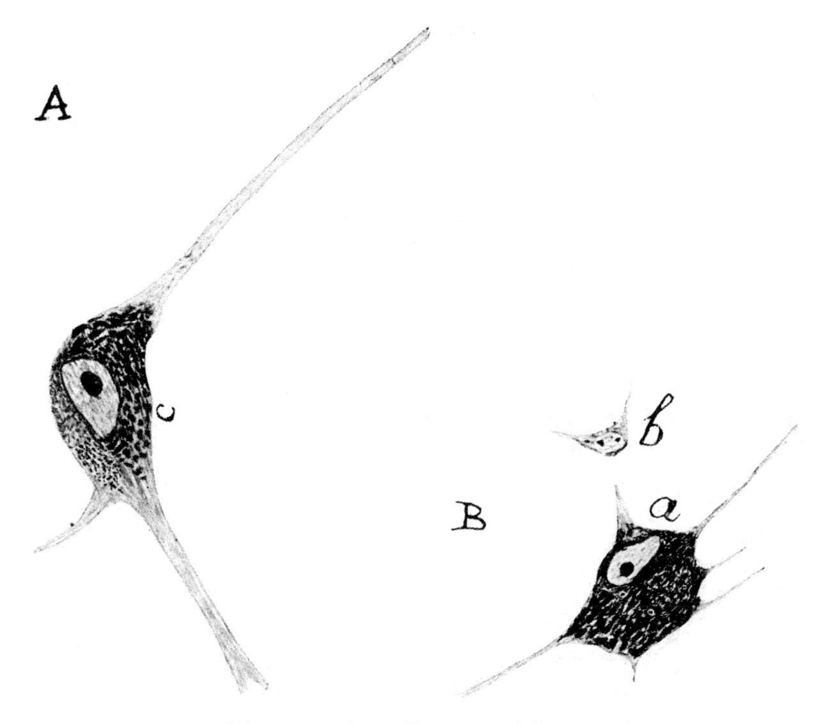

Fig. 4 A-c Large neuron of the STN, B-b small neuron of the STN (from Winkler 1928). For comparison, Winkler depicted a neuron of the rostral part of the nigra B-a

dendrites penetrate even the zona incerta, lateral hypothalamus and cerebral peduncle (Afsharpour 1985a, b). The nucleus is closed in cat (see e.g. Ramon y Cajal 1955), monkey and man. The dendrites are restricted to the nucleus, although some fusiform cells with their dendritic trees are found in the cerebral peduncle (Rafols and Fox 1976).

2.1 Neuronal Types Present in the Subthalamic Nucleus

In rat (Afsharpour 1985a, b; Kita et al. 1983) the neurons can vary from fusiform to oval or polygonal. Their imaginary circle diameters, from recalculated cross-sectional area, vary from 14 to 24 μm. In rat subthalamic organotypic cultures, the measured diameter from the cross-sectional area of the glutamate immunostained neurons is around 21 × 13 μm, and for parvalbumin-positive neurons 25 × 16 μm (Plenz et al. 1998). A volume of 0.8 mm^3 with 23,000 neurons was determined in the rat STN (Hardman et al. 2002). The soma does contain some somatic spines and produces 2–6 (Afsharpour 1985a, b) or 3–4 (Kita et al. 1983) primary dendrites. These dendritic stems were shown to produce long (500 μm) side branches that are sparsely covered with spines. The dendritic fields (100 × 600 × 300 μm; Hammond and Yelnik 1983) are in the sagittal plane oval and parallel to the long axis of the nucleus, while polygonal in the frontal plane. The axon originates from the soma and most axons can be followed out of the nuclear area, indicating that they are projection neurons. The axon bifurcates, giving off one branch to the globus pallidus via the cerebral peduncle, with the other branch ending in the substantia nigra. The pallidal axons give off branches into the entopeduncular nucleus. Of the studied STN neurons (Kita et al. 1983), 50% contained intranuclear axon collaterals. Since no Golgi type II neurons were found (Afsharpour 1985a, b), the main presence of Golgi type I neurons can be expected, which was already described earlier (Hammond and Yelnik 1983).

In the guinea-pig, three types of neurons can be discerned: multipolar neurons with 3–6 primary dendrites, bipolar neurons with two primary dendrites and pear shaped neurons with 1–2 dendritic trunks, arising from one pole of the neuron. The axon can arise from the soma or from the initial segment of a dendritic trunk (Robak et al. 2000).

In cat three types of neurons were discerned. Type I neurons are oval or polygonal with a cross-sectional area of 26 × 36 μm and four to six primary dendrites. Type II neurons are multipolar or polygonal, with a cross-sectional area of 31 × 43 μm containing four to seven primary dendrites. The type III neurons are polygonal with a cross-sectional area of 23 × 26 μm emitting four to six primary dendrites. Dendrites extended into the cerebral peduncle and cell bodies were found in between the peduncle fibres (Iwahori 1978). The subdivision in three subpopulations in cat was also supported by Romansky and Usunoff (1985). The large neurons constitute approximately 25%, the medium-sized about 50%, and the small neurons less than 5% of the neuronal population. The remaining 20% of neuronal perikarya do not offer reliable electron microscopic criteria that could allow an equivocal discrimination. It appears that from these 20% the majority represent medium-sized neurons, and none are small neurons (interneurons). These neurons are subdivided on their soma diameter: large neurons have a diameter above 18 μm, medium are in between 16 and 13 μm and small neurons contain a diameter below 12 μm. The serious difference in diameters between the results of Iwahori (1978) and Romansky and Usunoff (1985) depends on the difference in age and technique (i.e. kittens and Golgi were used by Iwahori; mature cats and electron microscopy by Romansky and Usunoff 1985). The cat nucleus tends towards a closed nucleus, since dendritic arborizations do not reach the lateral hypothalamus and zona incerta.

Larsen et al. (2004) stereologically studied the porcine STN. The STN neurons were medium-sized (diameter of 25–40 μm) possessing an oval to fusiform shaped cell body. Three to six dendrites originated from the soma, running parallel to the long axis within the nuclear area. The volume of the STN was 6.9 mm^3 and contained nearly 56,000 neurons that were glutamate positive.

In monkeys the species that have been studied are *M. mulatta*, *M. nemestrina* (Rafols and Fox 1976), *M. fascicularis* (Sato et al. 2000a, b) and the lesser bush baby (*Galago senegalensis*; Pearson et al. 1985). In the bush baby three types of neurons can be discerned, of which two are projection neurons, and the small soma neurons are considered interneurons. Similarly, in Macaca two types of projection neurons and one type of a local interneuron has been discerned (Rafols and Fox 1976).

The human STN has the largest volume and amount of neurons (240 mm^3 and 561,000 neurons, Hardman et al. 2002; 175 mm^3 and 239,500 neurons, Levesque and Parent 2005). Principal neurons in the human STN had a mean diameter of 24.6 μm, with an eccentrically located nucleus and nucleolus. The soma contained four principal dendritic trunks, with or without spines (Levesque and Parent 2005).

γ-Aminobutyric acid (GABA)ergic interneurons have been detected in the human STN (Levesque and Parent 2005). The smaller neurons that were glutamic acid decarboxylase (GAD)-positive had an ovoid cell body and a diameter of 12.2 μm. These interneurons produced two to three primary dendrites and are 70 μm long. "These dendrites were thin (1–2 μm), poorly branched, tortuous, and spread out in all directions. These interneurons contained lipofuscin" (Levesque and Parent 2005).

A comparative study involving rats, marmosets, macaques, baboons and humans (Hardman et al. 2002) has shown an increase in volume and neuron number from rat and marmoset to humans: rat 0.8 mm^3, 23,000; marmoset 2.7 mm^3, 34,000; macaque 34 mm^3, 154,000; baboon 50 mm^3, 229,000; man 240 mm^3 and 561,000 neurons. Yelnik and Percheron (1979) undertook another comparative study (see below).

Recalculations of the amount of cells per cubic millimetre STN from Hardman et al. (2002) show that the rat contains nearly 30,000 cells, while in man the amount is reduced to 2,300. Following Hardman's ordering of animal species, a constant reduction in the number of cells is evident (see Fig. 5).

An older comparative study on primates (Füssenich 1967) showed the same tendency for the row: Tupaia, lemur, Rhesus macaque, Pongo, Pan, gorilla and humans. The main difference between studies is the volume of the human STN, which is 67 mm^3 in Füssenich (1967), 175 mm^3 in Levesque and Parent (2005) and 240 mm^3 in Hardman et al. (2002); recalculated from Richter et al. (2004) it reaches 457 mm^3.

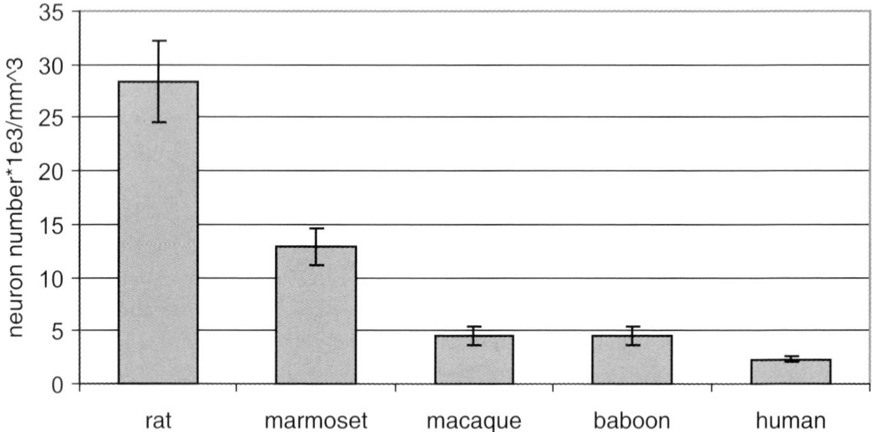

Fig. 5 The number of STN neurons per cubic millimetre. The phylogenetic row shows a steady decrease in the amount of STN neurons per cubic millimetre in which the human STN contains the lowest amount of neurons per cubic millimetre. (Recalculated from Hardman et al. 2002)

It is remarkable that in monkey and man interneurons are detected that are half the size of the projection neurons. Winkler's (1928) subdivision into small and large neurons (Fig. 4) is therefore correct. Moreover, he describes two subtypes of the large STN neurons, together with a topography (see start Sect. 2, this volume). In rat and pig a clear distinction in small and large neurons is missing. In cat at least one of the neuronal types belongs to the smaller neurons (diameter less than 12 μm, according to Romansky and Usunoff 1985).

In general two types of projection neurons are discerned, the third type of neuron presumably being the small interneuron. The projection neurons have (3) 4–6(7) primary dendrites and the interneurons 2–3. The dendrites of the projection neurons contain spines, while the dendrites of the interneurons are seemingly smoother, smaller and less long. In humans GABAergic interneurons have been detected (Levesque and Parent 2005). Whether these two types of projection neurons are identical to the two types of STN neurons that differentially react with low-threshold spikes or not, is unknown (see Sect. 3.2.2 of Part II of *The Subthalamic Nucleus*; Beurrier et al. 1999).

Contradictory to these results are those of Yelnik and Percheron (1979). In a morphometric study on Golgi-impregnated subthalamic neurons in macaque, baboon and man, compared to cat, only "Golgi type I neurons" could be detected. Moreover, "none of the small cells in baboon and in man had the characteristic morphology of a local circuit neuron". Studying these Golgi-impregnated neurons for dendritic numbers and for lengths, surfaces and 3D reconstructions, the average dendrite count was 7, giving 27 tips. The centrally located neurons showed ellipsoidal dendritic domains with mean dimensions of 1,200 × 600 × 300 μm.

2.2 Ultrastructural Features of Subthalamic Nucleus Terminal Boutons

The subthalamic neuropil contains a broad variety of terminal bouton types (Nakamura and Sutin 1972; Romansky et al. 1978, 1980a; Hassler et al. 1982; Romansky 1982; Chang et al. 1983; Romansky and Usunoff 1987; Bevan et al. 1995). According to the most comprehensive study in the cat (Romansky and Usunoff 1987), there are six distinct types of axonal endings: F1, F2, SR, LR1, LR2, and d.c.v. as well as "d" profiles—vesicle containing dendrites of the interneurons, participating in synaptic triads. The following description not only holds for the cat, but also for the baboon (Hassler et al. 1982; Usunoff et al. 1982a).

2.2.1 Flat Type 1 Boutons

Flat type (F1) boutons are the most commonly encountered in the STN neuropil and appear in almost each grid window. They usually arise from unmyelinated telodendria but occasionally they originate from myelinated axons with relatively thin myelin sheaths. The parent fibres give rise to a tandem of several F1 boutons, connected with slender intervaricose portions. The F1 terminals are elongated profiles with a length of 2.5–4 μm and a width of 0.5–1.3 μm. Occasional boutons are extremely extended, reaching a length of 8–9 μm. The F1 boutons contain a pleomorphic vesicle population, varying from almost round, through oval and slightly elongated, to markedly flattened synaptic vesicles; the latter accounts for 30% of the vesicle population. The F1 boutons form symmetrical axosomatic, axodendritic and axoaxonal contacts with the large and medium-sized projection neurons, but they never contact the interneurons. The axosomatic synapses cover most of the surface of the STN relay neurons (Fig. 6), and wedged F1 terminals between two perikarya, innervating both of them, is not a very rare finding. The action of some F1 tandems seems to be confined to a sole proximal dendrite and adjacent perikaryal surface, thus forming a powerful synaptic unit. Since the F1 boutons largely build the synaptic muffs around each proximal dendrite of the projection neurons, they might sieve off, as "filters", the other afferent impulses that reach more distal dendritic portions. More rarely, the F1 boutons contact somatic and dendritic spines. The fourth, and interestingly not a rare, postsynaptic target of the F1 boutons are the initial axonal segments of the large STN neurons. The F1 boutons in the cat are identical to the most common terminal (type 2) in the rat's STN, reported by Chang et al. (1983). The F1 boutons are the endings of the most powerful afferent connection of the STN, arising in the external pallidum (Romansky et al. 1980b; Usunoff et al. 1982b; and see below).

2.2.2 Flat Type 2 Boutons

Flat type 2 (F2) boutons are elongated profiles, arising from unmyelinated and, far more rarely, from very thin myelinated axons. Tandems of F2 terminals are observed

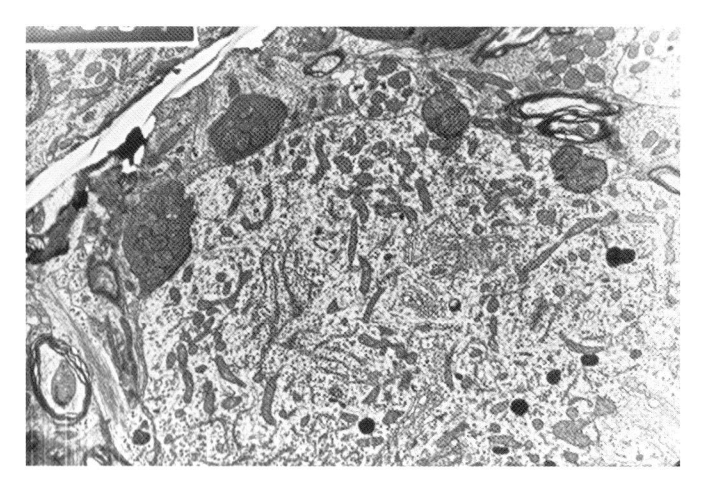

Fig. 6 Pallido-subthalamic degeneration at the ultrastructural level in the baboon (see Usunoff et al. 1982a)

less frequently than the commonly encountered by F1 boutons. They are also elongated, but their length rarely exceeds 4 μm. The vesicular population is pleomorphic, but the percentage of typically flattened vesicles is below 20%. The F2 boutons form symmetrical synapses, and the most common postsynaptic targets are the proximal dendrites of the large neurons. Although not as abundantly as F1, the F2 have been found to terminate on the somata of large and medium-sized neurons. They also occasionally contact the initial axonal segments. Importantly, unlike the F1, the F2 boutons contact the vesicle containing dendrites of the interneurons. The origin of F2 boutons remains unknown. Romansky and Usunoff (1987) speculated that they might represent terminals of the small local circuit neurons, but 20 years later this hypothesis is neither confirmed nor denied.

2.2.3 Small Round Boutons

The small round (SR) bouton type is relatively rarely encountered throughout the STN neuropil. The SR terminals arise, as a rule, from unmyelinated axons. Sometimes they are seen as two, rarely several, vesicle-filled axonal varicosities, connected by long conveying portions that are usually very thin (0.2–0.3 μm). The SR boutons are round or slightly elongated. Most of them measure 0.6–0.9 μm, and only rarely exceed 1 μm. On the other hand, some tiny exemplars measure only 0.4–0.5 μm. The vesicular population is markedly uniform, comprising small, round, tightly packed synaptic vesicles. The SR boutons form typical asymmetrical synapses on small distal dendrites and dendritic spines, while the contacts with larger dendrites are rare. Axosomatic contacts are practically absent, and the axoaxonal contacts with the initial portion of the STN axons are very few. On the other hand, the SR boutons fairly often take part in synapses with "d" profiles, and not so frequently with larger vesicle-containing dendrites of the interneurons.

At least the great majority of the SR boutons represent the endings of the corticosubthalamic axons (Romansky et al. 1979). There is some indirect evidence that they might represent collateral endings of the efferent axons of the STN neurons (Chang et al. 1984; and see Sect. 5, this volume).

2.2.4
Large Round Type 1 Boutons

Large round type 1 (LR1) bouton terminals were retrogradely traced to medium-sized or large axons of origin, having thick myelin sheaths. Sometimes robust fibres were found to break into several unmyelinated telodendria that give rise to individual LR1 terminals. By serial sections reconstruction, the LR1 boutons appear as several irregular or oval profiles, connected by kinky unmyelinated stalks that also contain synaptic vesicles. Such groupings of terminals appear as grape-like structures. The LR1 boutons exceed 2 μm, and the longest axis reaches 4.5 μm. The vesicular population comprises mostly oval, clear vesicles and a limited number of dense core vesicles. These terminals have multiple active zones and form asymmetrical synapses mainly on medium-sized and small dendrites and spines and, far more rarely, upon proximal dendrites and soma. Their constant participation in glomerulus-like formations is important. They are able to spread simultaneously their effects on a substantial number of dendritic postsynaptic sites, belonging to several neurons. In addition, the LR1 boutons form synapses with "d" profiles and vesicle-containing dendrites, thus influencing both relay neurons and interneurons.

The origin of the LR1 boutons is still unclarified. They are practically identical with cerebellofugal endings, but such a connection to the STN is unknown. Indications that such a connection could be present come from antegrade tracer studies in the hedgehog (Künzle 1998) and degeneration studies in the rat (Faull and Carman 1978), but these connections are still under discussion. They may represent endings of the thalamo-subthalamic tract (see Sugimoto and Hattori 1983; and Sect. 5, this volume).

2.2.5
Large Round Type 2 Boutons

The majority of large round type 2 (LR2) bouton terminals arise from unmyelinated axons, but serial sections reveal that very near to the point of termination the parent axons are medium-sized or even thick myelinated axons. Not infrequently the axon loses its myelin sheath immediately before an expansion of a LR2 terminal, followed by one-two terminals, interconnected by relatively short and thick intervaricose portions that also contain scattered vesicles. The LR2 boutons are oval or slightly elongated profiles that usually exceed 1.8 μm. As a rule they do not reach the dimensions of the LR1 boutons, and the largest exemplars display a longest diameter of 3.0–3.3 μm. The vesicular population comprises round and oval clear vesicles, slightly larger and more evenly distributed than in the LR1 vesicles. Single dense core vesicles are infrequently seen. The active zones are fewer than these of the LR1 boutons, but are usually more extensive. The LR2 boutons form asymmetric synapses. Along with contacts with thin dendritic shafts and spines, the LR2 boutons fairly often also contact the large, proximal dendrites, and the axosomatic synapses are more numerous than the LR1 axosomatic synapses. On the other hand, far less frequently they contact other vesicle-containing profiles.

A considerable percentage of the LR2 boutons in the STN represent the terminals of a connection arising in the pedunculopontine tegmental nucleus (Usunoff and Romansky 1983; Romansky and Usunoff 1983; Sugimoto and Hattori 1984), in agreement with several light microscopic studies (Rinvik et al. 1979; Nomura et al. 1980; Usunoff et al. 1982b; Hammond et al. 1983a; Jackson and Crossman 1983; Moon Edley and Graybiel 1983; Sugimoto and Hattori 1984; Woolf et al. 1990; Lavoie and Parent 1994a, b; Takakusaki et al. 1996; Ichinohe et al. 2000; Orieux et al. 2000). The LR2 bouton is also commonly observed in the substantia nigra (mainly in the zona compacta), and it also degenerates following destruction of the pedunculopontine tegmental nucleus (Usunoff and Romansky 1983; Usunoff 1984). It may be that a single neuron in this reticular formation nucleus gives rise to a branching axon that innervates simultaneously the substantia nigra and the STN. Moreover, by means of antidromic activations, Hammond et al. (1983a) demonstrated the existence of a branched axon from pedunculopontine nucleus to the STN and to the entopeduncular nucleus. These authors also demonstrated that the tegmento-subthalamic pathway in the rat is excitatory (in agreement with the ultra structural characteristics of the LR2 terminals) and estimated the rate of conduction of the pedunculopontine–subthalamic projection at approximately 1.7 m/s. Acetylcholine was suggested as a transmitter of pedunculopontine tegmental neurons by Shute and Lewis (1967), and later studies (Kimura et al. 1981; Mesulam et al. 1983, 1984; Satoh et al. 1983; Smith and Parent 1984; Sugimoto and Hattori 1984) reliably indicate this nucleus as a major cholinergic neuronal group. More recent studies, however, showed the pedunculopontine nucleus to include glutamatergic, GABAergic, peptidergic and dopaminergic neurons. Moreover non-cholinergic markers are co-expressed with markers of the cholinergic neurons, like glutamate, GABA and the NADPH (nicotinamide adenine dinucleotide phosphate, reduced) diaphoreses marker for nitric oxide (NO), and calcium-binding proteins (for overviews see Usunoff et al. 2003; Mena-Segovia et al. 2004).

2.2.6
Dense Core Vesicle Terminals

Dense core vesicle (d.c.v.) terminals are infrequently encountered in the STN neuropil. The d.c.v. boutons arise from unmyelinated axons that change "*en route*" their diameter, but remain generally thin (0.2–0.5 μm). They have a very tortuous course, and Romansky and Usunoff (1987) could never encounter a parent myelinated axon. The d.c.v. boutons contain oval or fusiform profiles and measure 1–2.2 μm. The vesicular population is pleomorphic and includes clear (agranular) and dense core vesicles. The clear vesicles are round and oval or, more rarely, elongated, while quite a few are markedly flattened. Along the "common" clear vesicles that measure approximately 42 nm, few clear vesicles are very large: 60–78 nm. The abundant dense core vesicles are the most typical feature of these terminals. They are round and oval, but can be elliptical, although that is rarer. As a rule, the dense core vesicles are large and exceed 80–90 nm, while the elliptical exemplars reach 120–135 nm. The more or less opaque core is surrounded by a translucent 6- to 8-nm ring. The d.c.v. boutons form mainly asymmetrical synapses. The active zones are single and small, and a long row of serial sections is often needed to recognize them. The most common targets of the d.c.v. boutons are small and medium-sized spines, followed by the dendritic spines. Quite more rarely they form synapses with large dendrites, and Romansky and Usunoff (1987) found in only two cases such boutons presynaptic to neuronal perikarya. These authors never encountered d.c.v. boutons in synaptic relationships with initial axonal segments, "d" profiles or vesicle-containing dendrites. The d.c.v. boutons appear to be of limited importance for the synaptology of the STN: they are very few in the baboon (Hassler et al. 1982) and in the cat (Romansky and Usunoff 1987). Apparently they are absent in the rat's STN (Chang et al. 1983). The origin of the d.c.v. boutons is unknown, but it was repeatedly speculated that they are monoaminergic terminals. Although by no means unique to such boutons, the large dense core vesicles were repeatedly suggested as their consistent feature (Hassler et al. 1970; Grofová and Rinvik 1971; Chan-Palay 1977; Buijs et al. 1984; Milner 1991 and many others). In the serotoninergic terminals this is especially true (Chan-Palay 1977, 1982; Groves and Wilson 1980; Beaudet and Descarries 1981; Wiklund et al. 1981; Pickel et al. 1984; Smiley and Goldman-Rakic 1996). There is also firm evidence that the large dense core vesicles often occur in peptide-containing axon varicosities and terminals (Pickel et al. 1979; DiFiglia and Aronin 1982; Somogyi et al. 1982a, b; Kapadia and de Lanerolle 1984; Smith et al. 1999; Waselus et al. 2005). Moreover, monoamine and peptide neurotransmitters or neuromodulators (or both types) may coexist within the same neuron, within its axon, and its terminals (Hokfelt et al. 1980; Johansson et al. 1981; Armstrong et al. 1984; Arvidsson et al. 1991; and references therein).

2.2.6.1 The Vesicle-Containing Dendrites of the Interneurons in the Subthalamic Nucleus

The "d" profiles often display irregular contours, or represent oval profiles with side protrusions. Their size varies widely, with some of them measuring about a micron, and the largest exceeding 5–6 μm. As a rule, the cytoplasm of the "d" profiles has the lowest electron density in comparison with all other vesicle containing profiles in the STN neuropil. The distribution of vesicles is very uneven. They are characteristically clustered and the adjacent areas are completely devoid of vesicles. The vesicular population is pleomorphic: from oval to flattened vesicles. The majority of the vesicles are elliptical. A constant feature of the "d" profiles is the occurrence of dilated cisternae of smooth endoplasmic reticulum, and—most important for the "diagnosis" of the terminal—free ribosomes. The "d" profiles possess presynaptic dense projections and form synapses of an intermediate type. The "d" profiles participate in synaptic triads (axo-dendro-dendritic synapses) as an intermediate component, and are postsynaptic to a variety of terminals: LR1, LR2, SR and F2. On the other hand, the "d" profiles are presynaptic to dendritic spines and "conventional" dendrites.

Despite a careful search, Chang et al. (1983) did not observe unequivocal "d" profiles participating in serial synapses. This coincides with the fact that the rodent STN lacks interneurons (Afsharpour 1985a), in contrast to the cat (Romansky et al. 1980a; Romansky and Usunoff 1985, 1987) and the monkey (Rafols and Fox 1976; Hassler et al. 1982; Pearson et al. 1985).

2.3 Cytochemistry of the Subthalamic Nucleus

The nucleus subthalamicus is an area with a high amount of capillaries in cat and rat (Friede 1966). The capillary length (in mm) per square millimetre is 1,152, which is high compared to the putamen (790) and caudate nucleus (770) and twice as high as in the substantia nigra. The capillaries are branches of the posterior communicating, posterior cerebral and anterior choroidal arteries (Parent 1996). The uptake of tritiated leucine (Altman 1963) and thio-amino acids in the mature rat, therefore, is high in the STN. Distant tracer injections of tritiated leucine contain the possibility of false-positive results.

The oxidative enzymes in the STN are markedly better represented than in the substantia nigra (Friede 1966). Cytochrome oxidase can be used for metabolic activity determination compared to baseline and is indeed high in the STN of MPTP-treated monkeys (Villa et al. 1996). The lysosomal enzyme acid phosphatase appears in the same density as its presence in putamen, caudate nucleus and substantia nigra. However, alkaline phosphatase is nearly twice as high in the STN as in other basal ganglion structures. It is presumably related to the high amount of capillaries.

In humans the amount of acetylcholinesterase as measured per wet weight in the STN and substantia nigra is low compared to the putamen and caudate nucleus. But acetylcholinesterase-stained sections show a strong species variation for the substantia nigra and STN, which restricts its use to a topographic tool for that species (see e.g. Paxinos and Huang 1995). In the rat, during development the acetylcholinesterase content in the STN is low and increases up to 4 months, after which a slight reduction is noticed towards maturity (Friede 1966).

The human STN contains consistent staining for iron, but less than what is present in the pars reticulata of the substantia nigra (Friede 1966). Iron plays an important role in the hypo-intensity of the STN on T_2-weighted images. Iron is present in the anterior medial part of the STN, but absent in its posterior levels. Therefore, the anterior part of the STN is hypo-dense and can be refound in the T_2-weighted images (Dormont et al. 2004). Copper can be detected in the STN in fair amounts, although the amounts in putamen and caudate nucleus are higher, with the substantia nigra being the highest.

The subthalamic neurons contain lipofuscin, which is a pigmented oxidized protein-lipid compound, stored in a granular form with autofluorescence capacity (Marani et al. 2006).

2.3.1
Nitric Oxide

The gaseous neurotransmitter NO is utilized by the STN neurons, especially in humans. Nisbet et al. (1994) found out that more than 95% of STN neurons are NO synthase (NOS) mRNA-positive. Eve et al. (1998) studied the expression of NOS mRNA in the basal ganglia of neurologically normal control subjects and patients with Parkinson's disease. In Parkinson's disease a significant increase in NOS mRNA expression was observed in the dorsal two-thirds of the STN with respect to the ventral one-third. Interesting age-related changes were reported by Cha et al. (2000). They reported that the number of NOS-immunoreactive neurons in the striatum and substantia innominata of the aged rat decreased, but the number of NOS-immunoreactive neurons in the STN increased in the aged rat.

2.3.2
Glial Fibrillary Acidic Protein

Glial fibrillary acidic protein (GFAP) is a marker for astrocytes or astrocyte activity (Goss and Morgan 1995). Astrocytes constitute nearly 50% of the brain's cells and envelope specific neuronal contact places called glomeruli (Ventura and Harris 1999; Grosche et al. 1999). Astrocytes have been demonstrated to be dysfunctional in various neurological disorders (see Seifert et al. 2006), and they are accompanied by astrocytic hypertrophy, an increase in astrocytic processes and an upregulation of the synthesis of GFAP (see Strömberg et al. 1986; Goss and Morgan 1995). An enormous astrogliosis has been shown in post-mortem tissue of Parkinson's patients (see e.g. Hirsch et al. 2003). A quantitative relation between neurons and glial cells has been described by Füssenich (1967). She found that the relation of 2.5 neurons to one glial cell is kept constant in the STN for man, and primates such as gorilla, Pan and Pongo. This relation increases to four to five neurons per glial cell in the primates rhesus macaque and Tupaia.

GFAP increase in astrogliosis has been found in the STN after iron injections in the striatum (Hironishi et al. 1999). 6-OH-DA injections into the striatum also produce GFAP glial activation (Henning J, Wree A, Gimsa J, Rolfs A, Benecke R, Gimsa U, in prep.). The explanation of such a distant glial reaction is proposed by Henning and colleagues: "Neuronal pathways could be accompanied by directional glial networks, which transmit activation signals if the neuronal pathway degenerates" (see also Sect. 3.2, this volume).

2.3.3
Ca^{2+} Binding Proteins

In the next volume (Part II) emphases will be laid on the channels for modelling STN behaviour in normal and parkinsonian neurons. Calcium channels play an important role in most cellular models proposed to date (see Sect. 3.2 of Part II of *The Subthalamic Nucleus*). Ca^{2+} binding proteins regulate intracellular calcium levels and therefore facilitating fast spiking activity (Augood et al. 1999) among effects on other functions. The three calcium-binding proteins (CaBPs): parvalbumin (PV), calretinin (CR) and calbindin-D28-k (CB) are distributed in the basal ganglia neurons according to a highly heterogeneous pattern (Parent et al. 1996; Hontanilla et al. 1998; Morel et al. 2002). In the rat's STN, PV-immunoreactive neurons and neuropil are concentrated in the lateral two-thirds of the nucleus, and STN is completely devoid of CB immunostaining (Hontanilla et al. 1997, 1998). Fortin and Parent (1996) compared the distribution of CR in the basal ganglia of the squirrel monkey, and described a high density of CR perikarya in the ventral tegmental area (VTA), moderate in SNc, low in SNr/SNl, and very low in STN, where it occurs only in specific sectors. In the squirrel monkey, the STN neurons are markedly enriched with PV but display only light CB immunostaining (Cote et al. 1991). The data on the human STN, provided by Augood et al. (1999), Morel et al. (2002), and Levesque and Parent (2005) are similar, albeit not identical. The three teams agree that the most common CaBP is the PV. According to Augood et al. (1999) the neurons in the dorsal STN are highly enriched, extending mediolaterally. Morel et al. (2002) declare that the most medial part of the STN is spared, and Levesque and Parent (2005) locate the PV-immunoreactive neurons in the dorsolateral STN. As in other mammals, the human STN displays a low CB immunoreactivity (Augood et al. 1999; Morel et al. 2002). The CR-positive neurons are located in the ventral STN according to Augood et al. (1999) with some overlap with the PV-positive neurons. Morel et al. (2002) locate them in the medial part, where the PV immunoreactivity is low, and Levesque and Parent (2005) found CR immunoreactivity in the ventromedial STN, delineated from the dorsolaterally located PV-positive neurons. Additionally, Levesque and Parent (2005) concluded that in the human STN the CaBPs are to be found in the projection neurons, but not in the GABAergic interneurons.

2.3.4 Receptors in the Subthalamic Nucleus

2.3.4.1 Dopamine Receptors

Dopamine receptors are classified into D_1 and D_2 receptor families and belong to the G protein-linked receptors. D_1 receptors are linked via adenyl cyclase to the cyclic AMP (cAMP) second messenger system. D_2 receptors inhibit the enzyme activity. Biological cloning techniques discerned five types of dopamine receptors: to the D_1 subfamily belongs the D_1 and D_5 receptor types, and to the D_2 subfamily the D_2, D_3 and D_4 receptor types. Although immunostaining for tyrosine hydroxylase showed a fine network of stained fibres, the presence of D_1 and D_2 mRNA-positivity in the human STN was absent (Augood et al. 2000). D_2–D_3 receptor mRNAs were found present in the human STN (Hurd et al. 2001).

Dopamine receptors are physiologically found functional in several species (see Cragg et al. 2004 and references herein). In the rat D_1, D_2 and D_3 receptor mRNA was found in the STN (Flores et al. 1999). Ligand binding and mRNA studies had already indicated such presence (Bouthenet et al. 1987, 1991; Boyson et al. 1986). Although binding sites for D_4 receptors could be distinguished, its mRNA was missing. It is supposed that the D_1, D_2 and D_3 can postsynaptically mediate dopamine, while D_4 receptors should be placed presynaptically (Flores et al. 1999). An extra argument for the presynaptic localization of D_4 receptors in the STN is given by the inhibition of GABA release by a selective dopamine D_4 receptor agonist in the absence of its mRNA in the STN (Floran et al. 2004). D_5 receptors are expressed in subthalamic neurons (Svenningsson and LeMoine 2002).

Autoradiographic studies with D_1 receptor antagonists, however, indicated that the D_1 family receptors are localized at "the ventral edge of the STN and dorsal aspect of the cerebral peduncle" (see Sect. 4.4.2, this volume; Kreiss et al. 1996). "A body of experimental results indicates that functional dopamine receptors are expressed in the STN but there is no agreement on the receptor types" (Baufreton et al. 2005); "The pattern of expression of DA receptor subtypes in STN neurons, examined through in situ hybridization, is controversial" (Cragg et al. 2004).

2.3.4.2
Cannabinoid Receptors

High levels of cannabinoid receptors are present in areas involved in the control of movement, especially the basal ganglia structures (Herkenham et al. 1991).

In general, cannabinoids inhibit neurotransmission by activation of potassium channels and inhibition of N- and Q-type voltage-gated calcium channels (see also Sect. 2.3.5, this volume). These effects are G protein mediated. Two types of cannabinoid receptors are discerned, CB_1 and CB_2. The CB_1 receptor is related to the nervous system, while CB_2 is involved in the immune system (except for the cerebellum). Cannabinoid receptor mRNA was found in the rat STN (Mailleux and Vanderhaeghen 1992). The CB_1 receptors of the STN are preferentially active presynaptically on axon terminals. To date no proof has been found that receptors are present on somatodendritic parts (Freiman and Szabo 2005). The immunohistochemical distribution of cannabinoid CB_1 receptors using antibodies did not substantiate the mRNA findings in the rat STN (Tsou et al. 1998).

Expression of cannabinoid receptors in the human STN is doubtful (Glass et al. 1997) and a clear change in cannabinoid receptors is not mentioned in the literature in the parkinsonian brain (Hurley et al. 2004).

2.3.4.3
Opioid Receptors

Of the series of opioid receptors, mainly the μ- and δ-receptors play a role in the STN. μ-Opioid receptors have been detected using mRNA expression methods in the human STN (Raynor et al. 1995). A decrease of this receptor has been noticed presumably in oral dyskinetic syndromes after chronic neuroleptic exposure (Sasaki et al. 1996; Shen and Johnson 2002). Zhu et al. (1995) found, using *hkor* mRNA, that the human STN contains a κ-like opioid receptor. Both μ- and δ-opioid receptors are expressed pre- and postsynaptically in the rat STN (Delfs et al. 1994; Florin et al. 2000). Opioids in the rat STN exert their inhibitory action on GABA release via both μ- and δ-receptors. The opioid receptors, in general, hyperpolarize neurons by increasing membrane potassium conductance and inhibit synaptic transmission by reducing voltage-dependent calcium currents (North 1993; see also Part II of *The Subthalamic Nucleus*).

2.3.4.4
Glutamate Receptors

Ionotropic and metabotropic glutamate receptors are found in the STN. The metabotropic glutamate receptors are G protein-coupled. From the three ionotropic subfamilies NMDA, AMPA, and Kainate, most receptors have been studied in experimental animals. A topological ligand-binding study in mice and their mutants showed the absence of all three subfamilies in the STN (Reader and Sénécal 2001).

The rat STN contains AMPA-Glu1 receptor, as detected by immunohistochemistry and in situ hybridization, AMPA-Glu2 receptor, found by in situ hybridization, and the following AMPA receptors: Glu2/3 and Glu 2/3/4 by immunohistochemistry; and Glu4 by immunohistochemistry and in situ hybridization. Moreover, NMDA receptor NR1 was detected with both immunocytochemistry and in situ hybridization in the rat STN (see Clarke and Bolam 1998, and references therein).

In the monkey STN, AMPA-Glu1 and -Glu2/3 receptors were found immunohistochemically on STN neurons and their proximal dendrites, albeit the Glu1 reaction was stronger. The metabotropic glutamate receptors mGlu1a and mGlu5 were found on dendrites and axon terminals. The NMDA-R1 receptor was represented on cell bodies and small-diameter dendrites. Ionotropic glutamate receptors were mainly found postsynaptically (Wang et al. 2000). The strong neuronal labelling for Glu1 was already confirmed by Ciliax et al. (1997). Huang et al. (2007) extended the presence of NMDA receptors with NR2D, NR2B and NR2C sub-units that contribute to the NMDA receptor channels.

"Binding studies indicate that NMDA, AMPA and kainate receptors are expressed at a low to moderate level in the STN in humans" (Smith et al. 2001; see also Lee and Choi 1992).

The metabotropic glutamate receptors are subdivided into three groups, I, II and III. The metabotropic glutamate receptors have been studied in the rat. Receptor agonists of group II and III produce behavioural effects, and immunocytochemical results support a function of these receptors in the rat STN (Kearney and Albin 2000). Blockade via haloperidol of group I, II and III metabotropic glutamate receptors produces effects in the STN of Parkinson's-induced rats (Miwa et al. 2000).

mRNAs in the rat subthalamic neurons were found positive for mGlu2 and mGlu3 (group II), while moderate labelling was found for mGlu5 (group I) (Testa et al. 1994). The mGlu2 localization was confirmed with mRNA by Ohishi et al. (1993). mGlu5 has been localized electron microscopically in dendrites of STN neurons (Awad et al. 2000).

In the human the metabotropic glutamate receptor type 2 (mGlu2) has been localized light microscopically in the STN, indicating that group II receptors are present (Philips et al. 2000).

In monkey STN, mGlu1a, of the group I receptors, showed localization into cell bodies, but cell localization was absent for mGlu5 receptors. However, positive labelling for both receptors of group I was found in the STN around axon terminals and dendrites (Wang et al. 2000).

In general, authors seem to favour perisynaptic localization for metabotropic glutamate receptors using ligand binding studies and postsynaptic localizations for ionotropic glutamate receptors. mRNA studies demonstrate neuronal localizations, indicating that these receptors can be present in the efferent axon terminals. "The functional role played by each of the particular subtypes is disputed" (Hamani et al. 2004).

2.3.4.5 GABA Receptors

GABA receptors are subdivided into GABA-A and -C, which are ligand-gated ion channel receptors, and GABA-B, a G protein-coupled receptor. Different GABA-A receptors are formed by assembly of multiple subunits. Subdivision of the GABA-A receptors is beyond the scope of this overview. Cloning of the GABA-B receptor delivered two subtypes named GABA-B_1 and GABA-B_2 that assemble in heterodimers to produce GABA-B receptors. GABA-B distribution was reported as negative in the rat STN (Bischoff et al. 1999; Durkin et al. 1999; Chen et al. 2004) using both ligand-binding and mRNA techniques. GABA-A-encoding genes were localized in the rat STN (Wisden et al. 1992; Zhang et al. 1991; Fritschy and Möhler 1995).

In monkeys, GABA-B1 immunoreactivity was present in the neuronal perikarya and dendrites. Negative and positive perikarya were evenly distributed throughout the nucleus. The neuropil was lightly stained. Electron microscopy showed GABA-B_1 localized at scattered pre-terminal axonal segments and axonal terminals (Charara et al. 2000). The GABA-B_2 subtype was also found to be expressed in monkeys (Billinton et al. 2000; Charara et al. 2000).

GABA-A receptors are highly expressed by mRNA techniques in the monkey's STN (Kultas-Ilinsky et al. 1998). Strong species variability is present in the subunits that constitute GABA-A receptors, especially between rat and monkey STN (Smith et al. 2001). The human STN displays both GABA-A and GABA-B receptors.

In monkeys and humans these receptors are expressed postsynaptically on dendrites and presynaptically in putative glutamatergic axon terminals in monkeys (Charara et al. 2000).

GABA-A is found at symmetrical synapses of the globus pallidus externus GABA-positive terminals in the STN (rat, Fujiyama et al. 2003; monkey, Galvan et al. 2004). GABA-B receptors were found in thalamo and cortical post- and presynaptic and perisynaptical areas (Boyes and Bolam 2007). In humans, a brain specific high-affinity GABA uptake protein (GAT-1) was found in STN neurons. "Accumulation of synaptically released GABA, via interaction with the GAT-1 GABA transporter, in the vicinity of their terminal projections is intriguing and may be of interest as a non-dopaminergic target for therapy in Parkinson's disease" (Augood et al. 1999).

2.3.4.6
Serotonin Receptors

Strong heterogeneity is present for the seven subtypes of 5-hydroxytryptamine (5-HT) receptors termed 5-HT-1 through 5-HT-7. Various binding sites could not be detected, although mRNAs have been present and vice versa. This makes a short overview difficult, and such a summing up is surely incomplete. 5-HT-1 has been demonstrated in the STN in the mouse [Mengod et al. 1990 (5-HT-1C)], the rat [Pompeiano et al. 1992 (5-HT-1A); Voigt et al. 1991; Bruinvels et al. 1993 (5-HT-1B); Wright et al. 1995 (5-HT-1C)], human [Waeber et al. 1989 (5-HT-1D), see their figures]. 5-HT-2 was found in the rat STN (5-HT-2C; Eberle-Wang et al. 1997, and references therein; Mengod et al. 1990; Pompeiano et al. 1994). In the monkey brain the STN was positive for the same mRNA (Lopéz-Giménez et al. 2001) as was displayed for the human STN (Pasqualetti et al. 1999). 5-HT-4 could be refound in the pictures from Vilaró et al. (2005) in the rat STN and for ligand binding in the guinea-pig STN. 5-HT-7 slight positivity was found in the human STN for this subunit (Martin-Cora and Pazos 2004). The presence of serotonin receptors in the mammalian STN is unequivocal (see also Sect. 5.3, this volume). Their subcellular structure, function and neurotransmitter interaction, however, are still unclear.

2.3.4.7
Cholinergic Receptors

Two types of cholinergic receptors have been discerned: muscarine, which is cascade-coupled, and nicotine receptors, a ligand-gated ion channel. For both types the subunit gene transcripts that constitute the receptors are known, but will not be considered in detail in this volume. For the nicotinic receptors the subunits $\alpha 4$, $\alpha 7$, and $\beta 2$ are most prominent in the mammalian brain. Muscarine receptors are subdivided into five subtypes m1–m5. However, only three subtypes [M1 (being m1, m4, m5), M2 (m2) and M3 (m3)] have been detected in vivo and can be G protein- or IP3-coupled in their effects.

In the rat brain using radioactive neuronal bungarotoxin binding, nicotine receptors were detected in the STN (Schulz et al. 1991). "Overall there is no obvious correspondence between the distribution of neuronal bungarotoxin binding sites and the presence of mRNA coding for any of the three α-subunits that have been characterized thus far" (Schulz et al. 1991). The immunohistochemical localization of the nicotine acetylcholine receptor demonstrated absence of positivity in the rat STN. Moreover, a discrepancy between these immunohistochemical results and bungarotoxin binding and radioactive nicotine labelling studies is evident. Presumably the antibody "does not recognize a protein that binds both radioactive-labelled nicotine and neuronal bungarotoxin" (Deutch et al. 1987). Serious doubt exists on the cholinergic action of nicotine receptors in the STN (Feger et al. 1979; Flores et al. 1996).

In monkey brains the STN contains the α4, α7 and β4 subunits of the nicotine receptor, as determined using in situ hybridization techniques (Quik et al. 2000), while Cimino et al. (1992) found an α3 subunit mRNA in the monkey's STN.

In humans, nicotine receptor binding studies showed a moderate activity in the STN and a decrease in Parkinson's disease (Pimlott et al. 2004).

In rats the localization of m3 receptor protein and M3 receptor binding in the STN is unequivocal (Zubieta and Frey 1993; Levey et al. 1994). In the STN, expression occurred for m3 and m4 mRNA (Weiner et al. 1990). However, the expression for mRNA for the M3 receptor is weak in the STN (Shen and Johnson 2000). Muscarine receptor M3 is kept responsible postsynaptically for the cholinergic activation of STN neurons (Flores et al. 1996; Shen and Johnson 2007). "Curiously, to my knowledge, no new reports of functional studies on human brain muscarinic receptors have appeared in the last decade" (Raiteri 2006).

2.3.5 Ca^{2+} Channels

Calcium channels are subdivided in a low voltage-activated channel (T-type) and several high voltage-activated channels (L-, N-, P-, Q- and R-types). Most of the localizations are demonstrated by physiological methods. Genetically determined subunits that constitute these channels have been described. Calcium channels are multi-subunit complexes and are mainly studied in the rat. The neuronal calcium channel α_{1E} subunit was found with immunohistochemistry in the neurons of the rat STN (Yokoyama et al. 1995). The α_{1A}, α_{1B}, α_{1C} and α_{1D} subunits are not mentioned as being present in the rat STN (Tanaka et al. 1995). The α_1 subunits are thought elementary for the high voltage-activated calcium channels if combined with other subunits. On its own, α_1 subunit assembly gives a T-type similar channel (Meir et al. 1999). The α_{1E} subunit is thought to be involved in structuring channels similar to low voltage-activated channels, presumably the R-type. The subunits α_{1G}, α_{1H}, α_{1I} that determine T-type calcium channels are all three present in the rat STN, with high labelling for the α_{1I} unit (Talley et al. 1999).

The α2δ1, α2δ2 and α2δ3 subunits of calcium channels were also demonstrated in the rat STN by mRNA techniques (Cole et al. 2005). α2δ2 expression was found high in the caudal rat STN. These subunits are thought to be related to the L-type, non-L-type and T-type calcium channels, provided they combine with one of the variations of the α1 subtypes (Cole et al. 2005).

In the rat, several of the subtypes of high voltage-activated channels (L-, T/R-types), were demonstrated by electrophysiological techniques (Beurrier et al. 1999; see also Sect. 3.1.2 of Part II of *The Subthalamic Nucleus*). A preferential localization of the low voltage-activated channel (T-type; see also Sect. 3.3 of Part II of the next volume) in STN dendrites was favoured (Song et al. 2000) and L-type in the soma (Otsuka et al. 2001). The results of colocalization experiments, mainly in rats, demonstrated that many nerve terminals, including those in afferents and efferents of the STN, possess more than one type of calcium channel involved in transmitter release (see Meir et al. 1999).

Glutamate release is dominated by P/Q-type calcium channels (Turner et al. 1993). However, the rat STN contains some low-voltage and, except for the P-type, all of the high-voltage activated channels (Song et al. 2000; see Sect. 3.3 of Part II of *The Subthalamic Nucleus*).

The confusing results obtained with mRNA methods ask for extensive research. However, Ca^{2+} channels are present in neurons and dendrites of the rat STN. T-type calcium channels are also found in the human as indicated by mRNA of the $\alpha 1$ subunit (Monteil et al. 2000).

2.3.6
Purinergic Modulation

Recently the adenosine receptors have been found to be active in the basal ganglia. In the STN, adenosine receptor A_1 has been found (see Misgeld et al. 2007). Consumption of caffeine should be neuroprotective and is expected to reduce the loss of dopamine in Parkinson's disease and does so by blocking the A_{2A} receptor in MPTP-treated animals (Chen et al. 2001). The presence of this A_{2A} receptor in the STN is not reported. Adenosine is produced from AMP by the enzyme 5' nucleotidase. The enzyme cannot withstand fixation. Convincing 5' nucleotidase ultrastructural localizations are therefore mostly absent in literature (see Marani 1982) and localizations of adenosine production/storage sites are hard to compare to receptor sites. Effects of purinergic modulation are described by Kitai (2007) for the globus pallidus externus. "Increase of extracellular GABA levels after lesion of the nigrostriatal connections to the striatum and globus pallidus externus is blocked by A_{2A} antagonists". Activation of the receptor increased the GABA release after electric stimulation of GABA connections.

3
Ontogeny of the Subthalamic Nucleus

3.1
Development of the Subthalamic Cell Cord

During and after the formation of the neural tube the ventricular surface contains a series of ridges and sulci. This pattern of ridges and sulci demonstrates a segmental pattern. The ventricular surface morphology has been studied with SEM in the diencephalon of the rat, and subsequently the developmental changes were described as to its neuromeric borders and the presence of the sulcus limitans (see Lakke et al. 1988).

Cytological studies and ^{3}H-thymidin autoradiography in the Chinese hamster revealed the origin and development of the nucleus subthalamicus (Keyser 1972). Cytological studies by Richter (1965) and Müller and O'Rahilly (1988a, b, 1990) demonstrated the early development of this nucleus in man. This description follows the results in the Chinese hamster. Chinese hamster borders were compared previously to the borders in the rat (see Lakke et al. 1988) and will be compared in this description to those obtained in man.

The regio subthalamica in the Chinese hamster develops rostral of the synencephalon (future regio pretectalis) and basal to both the pars dorsalis (parencephalon posterior) and ventralis thalami (parencephalon anterior) (for localization of these parts on the ventricular surface see Fig. 7). At E_{12} (embryonic day 12) the neuromeres are clearly discernable. They show that the basal parts of the future regio pretectalis, the area below it (the future regio subthalamica), and the area caudal to the eyestalk are advanced in development compared to the other ventricular areas. At E_{13} a mantle layer can be discerned in these areas. A compact "stream" (Keyser 1972) of cells can be noted and is called the subthalamic cell cord. This cell cord does not contain condensations or a development that indicates pronuclei. The cord reaches from its basal part below the future main thalamic (parencephalic) areas towards the diencephalic telencephalic border that can be recognized by the external telodiencephalic groove (also called the sulcus haemisphericus). So, early in development (E_{12}-E_{13}) a subthalamic region can be discerned behind the mammillary recessus and a subthalamic cell cord extending from this area, reaching to the telencephalic-diencephalic border.

At E_{14} the prerubral tegmentum contains the pronuclei of the nuclei of Cajal and Darkschewitsch and it continues without clear borders in the more rostrally situated regio subthalamica. The basal area of the subthalamic cell cord is slightly more developed than the rest of this cell cord. It is this stage of development in which the growth of the mantle layer obscures the neuromeric borders. However, the neuromeric subdivisions can still be discerned in the basal/caudal diencephalic part, delineating the subthalamic area and the start of its cell cord. It should be noted here that the basal regio subthalamica is sandwiched between the tegmentum prerubralis and the regio mamillaris, while from its rostral part the subthalamic cell cord rises towards the diencephalic–telencephalic border.

At E_{15}, within the regio subthalamica develops the supramammillary commissure that crosses the midline. The development of the regio subthalamica is advanced, and it is proposed that this area is "an organizer centre for the rostralmost parts of the brain" (Keyser 1972). The corpus subthalamicum Luysii (the future nucleus subthalamicus) is recognizable at this stage. "It develops from a tangentially migrating stream of cells that, originating from the matrix of the supramammillary recess, gradually shifts in a rostrodorsal direction" (Keyser 1972).

At E_{16} the subthalamic cell cord has developed into the suprapeduncular complex and a superficial corpus subthalamicum can be recognized. The suprapeduncular complex contains several subdivisions, but seemingly contributes to the entopeduncular nucleus and the globus pallidus (the rostral extension of the suprapeduncular complex). Translated into human terminology, it contributes to the globus pallidus internus and externus, respectively. A posterior hypothalamic area is considered to develop in relation to the corpus subthalamicum.

The future corpus Luysii has a latero-dorsal migration, while the remnants of the subthalamic cell cord contain a more medio-dorsal development, in fact

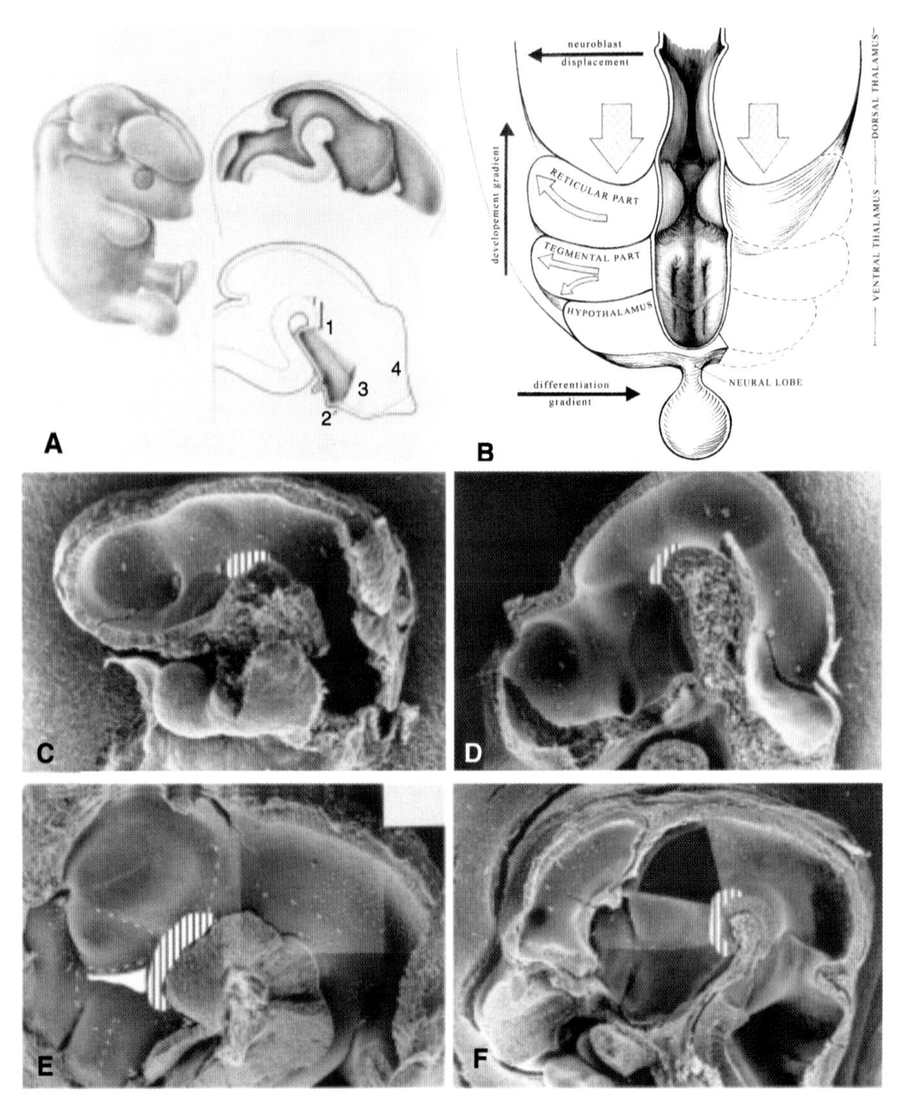

Fig. 7 A Drawing of a rat embryo showing the localization of brain and brainstem shining through the skull. Depicted is the luminary surface and next to it the hypothalamic and subthalamic region. *1*: mammillary region, *2*: chiasmatic area, *3*: optic stalk, *4*: future foramen of Monroi. B A late developmental stage in a transverse section is shown: neuroblast displacement from the ventricular matrix towards the pial surface, the opposite differentiation gradient and the developmental gradient. The tegmental (subthalamic) part shows the two types of migrations: medio-dorsal and latero-dorsal. C (E_{11}), D (E_{12}), E (E_{13-14}), F (E_{15}): Successive stages in the development of the rat ventricular surface. *Grey shadowing* is the hypothalamic area and *striped* is the subthalamic area. The border to the prerubral tegmental area is difficult to establish. In E the sulcus limitans, borders of the parencephalon dorsalis and ventralis, and the border of the hypothalamic area are *stippled*. In E is also depicted in *white* the subthalamic cell cord. *Arrows* indicate the recessus opticus and mamillaris. F *Striped* is the subthalamic area, in *grey shading* the parencephalon dorsalis and the hypothalamic area. The parencephalon ventralis can be recognized between the two areas

separating both cell populations. Their base is already loosened at E16. Thus, the subthalamic area really undergoes a tangential migration, in which the corpus Luysii takes the most lateral route.

The autoradiographic results in the Chinese hamster show that the area subthalamica is characterized by an early onset of differentiation. It is autoradiographically labelled from E_{12} until E_{18} onwards, being one of the first labelled areas. It originates from the medially located ventricular surface of the area above the mammillary recess and spreads tangentially. The most laterally situated cells of the corpus Luysii are born on E_{13} to E_{18}. The tangential migration is also supported by the autoradiographic results.

In conclusion:

1. The regio subthalamica is one of the first to differentiate in the ventricular surface.
2. The regio subthalamica gives rise to the subthalamic cell cord. The regio subthalamica contributes to a posterior hypothalamic area.
3. The regio subthalamica produces in its basal ventricular matrix the corpus Luysii and the suprapeduncular complex.
4. The matrix is localized just above the mammillary recess also called the supramammillary recess.
5. The corpus subthalamicum Luysii takes a more lateral direction or tangential migration than the suprapeduncular complex does.
6. The border between the subthalamic area and prerubral area is difficult to discern.

In rat, Marchand (1987) used autoradiography and determined the germinative zone just caudally and dorsal of the mammillary recessus as the origin of the STN. The first migration is oriented radially and thereafter tangentially. These results are well comparable to those of the Chinese hamster.

The gene expressions in longitudinal strips for the developing diencephalic and forebrain areas in fact show that the subdivisions as determined by Keyser (1972) are generally correct. The hypothalamic area is characterized by Nkx-2.1, the hypothalamic cell cord by both Nkx-2.1 and 2.2, while the supramammillary region with the rubro-tegmental area is characterized by Shh (Shimamura et al. 1995). The expression of Dlx1/2 and Wnt-3 also confirm the main boundaries of the hypothalamic cell cord and the rubro-tegmental area (Bulfone et al. 1993).

The short description of Hamani et al. (2004) that the hypothalamus sensu lato is further subdivided into a hypothalamus sensu stricto and subthalamus (taken over from Müller and O'Rahilly 1988a, b, 1990) is not substantiated by the results in rat, Chinese hamster and man (Richter 1965). In the ventricular surface a strict subdivision in hypothalamic cell cord versus subthalamic area and subthalamic cell cord can be made (see Figs. 7 and 8 for the consequent subdivision in subthalamic and hypothalamic areas).

The development of the subthalamic region in man has been earlier described by Barbè (1938). The first recognition of the subthalamic area is at 83 days of development. Barbè includes the mammillary bodies in this region. The corpus Luysii can be discerned at 92 days of the human brain development. The extensive studies of Müller and O'Rahilly (1988a, b, 1990) on the development of the human neural tube

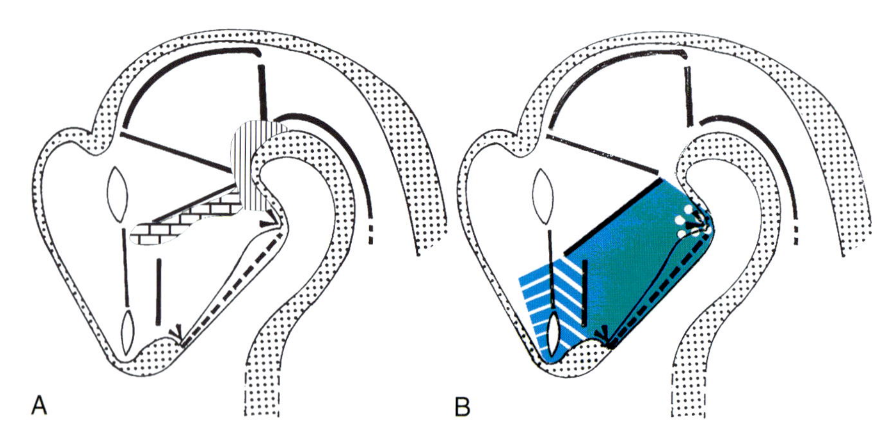

Fig. 8 A Schematic drawing of both the subthalamic area (E_{16-18}, *striped*) and the subthalamic cell cord (E_{13-14}, *blocked*) in the *upper part* of the figure as compared to the hypothalamic area at E_{18} in the *lower part*. B The hypothalamic area is subdivided in a suprachiasmatic area (*striped*), an infundibular (*grey*) and a mammillary (*circles*) part. Overlap of the hypothalamic and subthalamic area is present in the posterior hypothalamic region. *Vs* indicate, from *left to right*, recessus opticus and recessus mamillare

from stage 13 (28 days) to 21 (53–54 days)–23 (56–60 days) demonstrates the subthalamic area far earlier (from stage 13 onwards) as part of the caudal synencephalon and parencephalon (caudal D_2 in Müller and O'Rahilly 1988a, b). However, the origin of the subthalamic area is not given at this stage. A subthalamic area is described just parallel and above the hypothalamic cell cord, which is identical to the subthalamic cell cord of Keyser (1972). The further development of the STN is above the mammillary recess, adjacent to the mammillary bodies near the mesencephalic border at stage 18 (45 days). The migration stream for the STN is not evident.

The tegmental-rubral area develops early as was also found in the Chinese hamster. A border between the mesencephalic and subthalamic area was also not given by Müller and O'Rahilly (1988a, b, 1990).

In man the subthalamic cell cord is held responsible for the production of the corpus subthalamicus Luysii (second month of development/18 mm; Richter 1965), and its tangential spread is presumably later in development (fourth month, 115 mm; Richter 1965). Müller and O'Rahilly (1988a, b, 1990) and Richter (1965) place the development of the STN above the mammillary recess, and Müller and O'Rahilly (1988a, b, 1990) far earlier (45 days) than Richter (1965), who claims that in earlier stages the development is still in the matrix layer and difficult to follow, without indicating a tangential migration stream. The difference between the Chinese hamster and man, therefore, is not the diencephalic part where the STN arises: supramammillary recess (Richter 1965; Müller and O'Rahilly 1988a, b, 1990) in fact confirms the results of the Chinese hamster. The probability that an early tangential migration of the STN is present cannot be determined in human material; a late migration is advocated by Richter (1965). Autoradiography cannot be performed in man; therefore one should consider that the Chinese hamster results are well based, since results in man can only be studied by differential appearances of cytological characteristics in matrices. Richter (1965) indicates in his study of the 37-mm embryo (third month) that based on the cellular development of the STN a ventrolateral (upper) and a dorsomedial (deeper) part can be discerned.

3.2 Early Development of Subthalamic Connections

The early development of the fibres in the striatum can be followed using myelin colouration. It was Flechsig (1876) who detected that certain parts of the striatal complex were advanced in their myelinization. Using the Pal staining or other myelin colourations, these early myelinating bundles can be traced. Central nervous system areas can be distinguished by the combination with a cell staining, mainly carmine red or Gieson green in those early days. Inherently the idea is that "what myelinates first developed first".

Kodama (1927) using haematoxylin eosin and the Pal/carmine technique described 50 human foetuses and postnatal babies. In this monograph we will follow Kodama's overview.

The earliest result in a 5-month-old foetus is the myelinated connection between the globus pallidus and the STN. The internal part gives off more myelinated fibres than the external part, both projecting into the STN. Kodama subdivides the STN in a medial (parvo-) and lateral (magnocellular) part. The lateral part differentiates earlier than the medial part. The lateral part receives the earliest connections: the pallido-subthalamic connections. It can be noted in Kodama's figures that these fibres constitute the dorsolateral division of Edingers "comb system". The medial part contains connections at the seventh foetal month that can be followed into the supramammillary commissure towards the contra lateral STN.

Therefore, from the earliest studies it can be concluded that the magnocellular lateral part is involved in the pallido-subthalamic connections, while the parvocellular medial part is related to the connections between the ipsi- and contralateral subthalamic nuclei.

Richter (1965) also shows the early myelinization of the pallidum and nucleus subthalamicus. If the subthalamic cell cord indeed contributes to the pallidum (see rat, Ströer 1956; man, Spatz 1925, 1927; Richter 1965; rabbit, Rose 1942, and others, see above) and the migration/differentiation stream of the subthalamic cell cord originates from the same area in which the nucleus subthalamicus develops (later on), then it could well be possible that the axonal connections do grow out accordingly. Arguments for this hypothesis can be found in Kordower and Mufson (2004) in which NGF receptor immunoreactivity is equally (timely) expressed in both the pallidal and subthalamic regions. Since the rubral-tegmental area is difficult to separate from the subthalamic area, one should be aware that subthalamic-nigral connections are homologous to the pallido-subthalamic connections that are probably imprinted during development.

As mentioned above, the development of the tracti in the Chinese hamster follows the description of Keyser (1972). At E_{14} the stria medullaris and tractus opticus are present and recognizable in the rostral diencephalon. However, at the same age the fasciculus mammillo-tegmentalis is present, just located below the subthalamic cell cord and within the subthalamic area. Within the area above the subthalamic cell cord the tract of the zona limitans is present.

The supramammillary commissure is developed at E_{15}. At E_{16} the fasciculus mamillo-tegmentalis is reaching the area of the lateral rubral tegmentum, and the fasciculus retroflexus reaches from the substantia nigra towards the habenulum. From this age on the main tracti are discerned and extend their presence.

Gene expression studies have also looked at the connections from forebrain towards the mesencephalon. The results for the optic tract (Tuttle et al. 1998) and medial forebrain bundle (Martin et al. 1985) support the original results of Keyser (1972).

4 Topography of the Rat, Cat, Baboon and Human Subthalamic Nucleus

The cytoarchitecture, rostral and caudal borders of the mesencephalon, neighbouring nuclei and tracts and the general topography of the STN are exemplified here by typical species, namely rat, cat, baboon, and man. The STN is e.g. considered a "biconvex-shaped structure surrounded by dense bundles of myelinated fibres" (see Hamani et al. 2004), consolidating the idea of a well-bordered nucleus. Rostral and caudal borders of e.g. the human STN are indistinct. Therefore, extra Nissl and Weigert stained sections of these areas are added to the human series. These human sections coincide with those of the *Atlas of the Human Brain* (Mai et al. 1997) by number. Neighbouring nuclei and tracts, therefore, can be localized easily using this atlas, which has its direction well defined.

4.1 The Rat Subthalamic Nucleus: Cytoarchitecture

Going from caudal to rostral, the first rat subthalamic cells are found at the ending of the substantia nigra, within the dorsal part of the cerebral peduncle (Fig. 9, 495), both in Woelcke (1942) and Nissl stainings (see Voogd and Feirabend 1981). The next level shows grouping of the subthalamic neurons just in the middle of the cerebral peduncle. Some cells are placed somewhat more inwardly (Fig. 9, 501). The nucleus develops more rostrally over the whole peduncle, containing medially more neurons than laterally; the head being 5–6 cell layers thick, its lateral cauda containing one to two cell layers (Fig. 9, 507). Due to the lateral extension of the cerebral peduncle towards rostral, the cauda breaks up in separate islands, while the head grows and occupies more than half of the cerebral peduncle. At a medial quarter of the cerebral peduncle 10–12 layers of subthalamic neurons can be discerned (Fig. 9, 513). At the medial side of the cerebral peduncle subthalamic

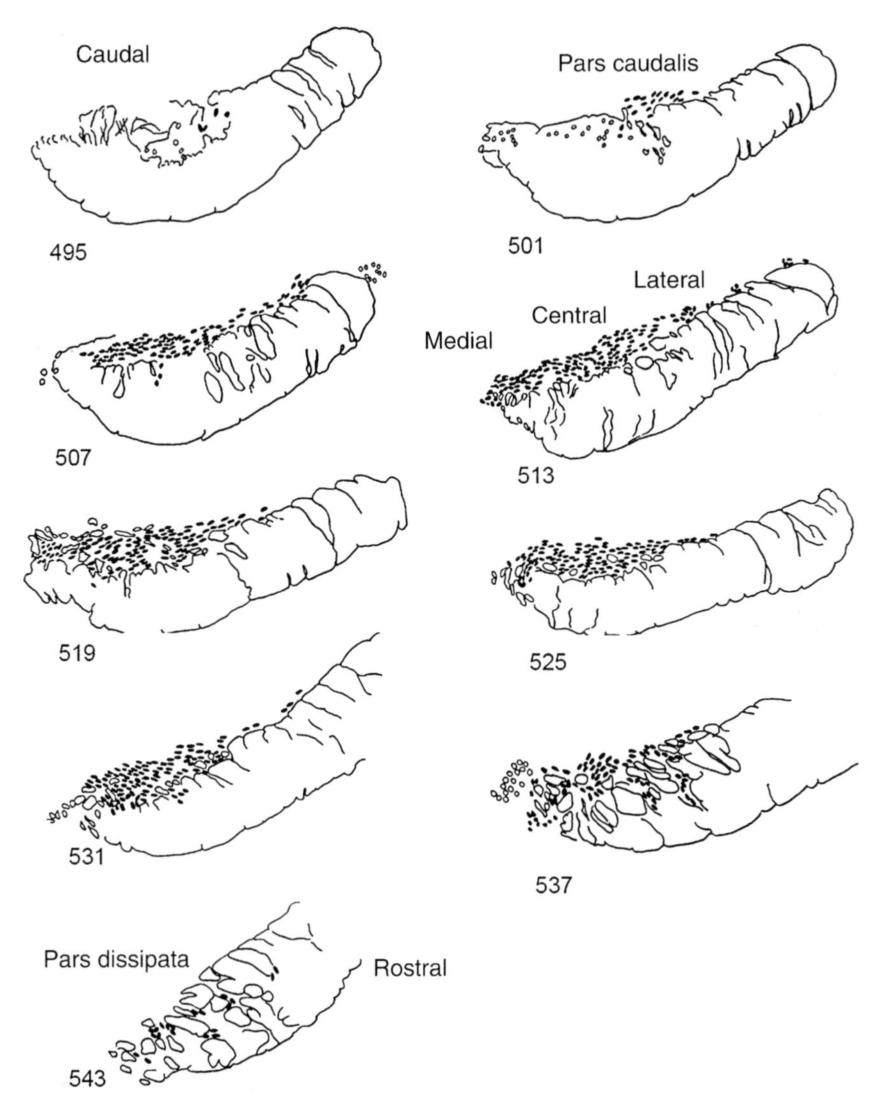

Fig. 9 Camera lucida drawings of transversal Nissl sections of the rat STN. Neurons are depicted in an approx. 1:3 ratio

neurons curl around the peduncle end superficially. The cauda, one to two cell layers thick, hardly extends over the medial half of the cerebral peduncle (Fig. 9, 519). A group of neurons with smaller cells seems to concentrate at the medial end of the cerebral peduncle. The head of the STN nucleus now reaches its elliptic form, with a very short cauda of few neurons. Bundles of the cerebral peduncle are intermingled with neurons and neuropil. The peduncle now extends more medially than do the subthalamic neurons (Fig. 9, 525). The nucleus is more rostrally restricted to the medial one-third of the cerebral peduncle. Only a few single cells of the cauda can be perceived in the peduncle lateral two-thirds. The head of the nucleus is flattened and more cells are encountered in between the superficial peduncle bundles (Fig. 9, 531). At the rostral end of the nucleus the subthalamic neurons are localized in between the superficial and deeper peduncle bundles (Fig. 9, 537, 543). It is this level where a group of round neurons, not belonging to the STN, are localized against the medial end of the cerebral peduncle. The STN ends by dispersed neurons in between the medially localized peduncle bundles (Fig. 9, 543).

Based on this description the rat STN can be subdivided into several parts (see Fig. 9). The caudal start can be recognized as a separate entity, called pars caudalis, which continues into a pars medialis, pars centralis and a pars lateralis. The pars centralis is recognized by its concentration of neurons, while the pars lateralis comprises a small layer of cells over the cerebral peduncle.

Tapering on both sides of the rostral ending of the pars centralis is found the pars dissipata present in between the peduncular bundles.

4.2 The Cat Subthalamic Nuclear Area: Sagittal Topographic Borders

In Fig. 10, sagittal alternating Woelcke (1942) and Nissl sections (Voogd and Feirabend 1981) of the mesencephalon of the cat are shown (Usunoff 1990). These sagittal sections of the cat mesencephalon and diencephalon demonstrate how difficult it is to border mesencephalon and diencephalon in the cat's mature brain. According to Voogd in Nieuwenhuijzen et al. (1998), this border at the end of the tegmentum mesencephali passes rostral of the substantia nigra. "The pes mesencephali comprises the cerebral peduncle and the substantia nigra. The medial lemniscus marks the dorsal border of the pes mesencephali. Its caudal border is determined by the rostral border of the pons and the lateral by the rostral margin of the brachium pontis". In the Nissl sections, the STN (Fig. 10) can easily be recognized by its dense packing of cells. Moreover, the intimate relation between STN and the substantia nigra is present. Rinvik's cerebral peduncle loop (1968; tractus corticotegmento-thalamicus Rinviki, Usunoff 1990) is present in between both structures in the direction of the VPL (nucleus ventralis posterior lateralis thalami). However, this bundle is specific for the cat. This bundle is situated at the mesencephalic–diencephalic border. The zona incerta tapers rostrally over the STN, to cover it over its whole extent with its main concentration of neurons. The zona incerta's rostral and caudal poles do have different connections compared to the central part (Romanowski et al. 1985). Although the central part is reported to be involved in visuomotor integration, feeding, drinking and locomotion, a relation with the basal ganglia or with the STN has not been reported (see Voogd in Nieuwenhuys et al. 1988). Field H2 of Forel covers the STN and contains neurons

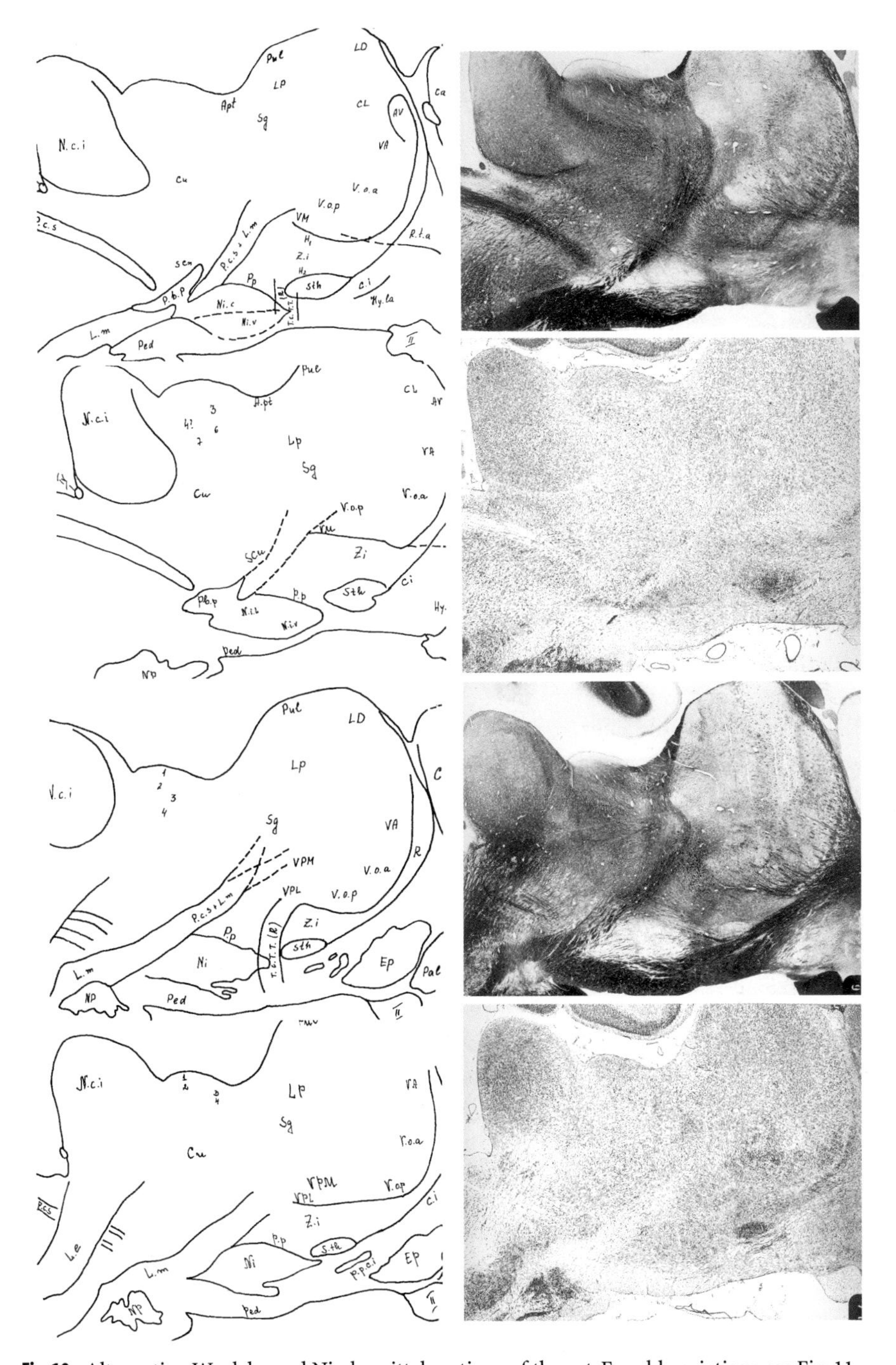

Fig. 10 Alternating Woelcke and Nissl sagittal sections of the cat. For abbreviations see Fig. 11

that are few. The nature of these neurons is unclear; whether or not they belong to the STN or the zona incerta is unknown.

The intimate relation between the capsula interna/pedunculus cerebri and the STN is clearly present in the Woelcke sections, as holds for the substantia nigra. The medial lemniscus presents itself at the caudal pole of the substantia nigra, goes obliquely to dorsal and over-roofs the area of the zona incerta and STN. The intimate relation between capsula interna/pedunculus cerebri with the lateral hypothalamus and pallidum, and the STN just above it, can be noticed.

4.3 The Baboon Subthalamic Area: Nuclei and Tracts

The baboon STN has been presented in a series of sections by Usunoff (1990) for the description of the topography of the substantia nigra (distance between transverse sections is 380 μm in the alternating Woelcke and Nissl order, Fig. 11). The intimate relation between substantia nigra and STN is noted, since the start of the STN is difficult to discern from the lateral part of the substantia nigra (Fig. 11.1 Woelcke sections). Field H2 of Forel does finely separate the STN from the zona incerta, while the start of the nucleus is just below the central, medial part of the zona incerta.

In the next Nissl section (Fig. 11.2) it is clear that the pars reticulata borders the STN. Its medial end points into the pars compacta. The whole STN has not yet a relation to the capsula interna, only its lateral part does. The STN is easily recognized by its high concentration of cells. In the next Woelcke section a plate of white matter intermingles between the substantia nigra and the STN, separating both nuclei partially. Here the medial edge of the STN is related to the pars compacta of the substantia nigra, indicating that the lateral part of the pars reticulata is firstly separated from the STN.

The relations of the STN with the substantia nigra are easy to detect. The STN's medial edge is placed next to the pars compacta, while the lateral edge is related to the capsula interna. The rest of the medial half of the STN is placed near to the pars reticulata. Above the STN the increase of field H2 is noted. It looks as if the increase of the STN forces the zona incerta into its cauda. The globus pallidus is nearing the STN, always separated from the STN by the capsula interna. The STN reaches its maximal extent in Fig. 11.4 and 11.5. The STN is now separated from the substantia nigra by a layer of fibres, while at its dorsal side the field H2 of Forel has reached its maximum thickness. The capsula interna proceeds towards medial and the zona incerta diminishes. The first indication of a redistribution of the concentration of cells is noted: the lateral part of the medial half contains a lower amount of cells. The pictures of Fig. 11.5 show the complete extent of the STN. The Nissl section allows a tripartition of the STN into an oblique medial part with a high concentration of cells, an oblique middle part with a low concentration of cells and an oblique lateral part with a high concentration of cells. The medial half of the STN is lying above the pars reticulata of the substantia nigra, while its medial point is just above the pars compacta.

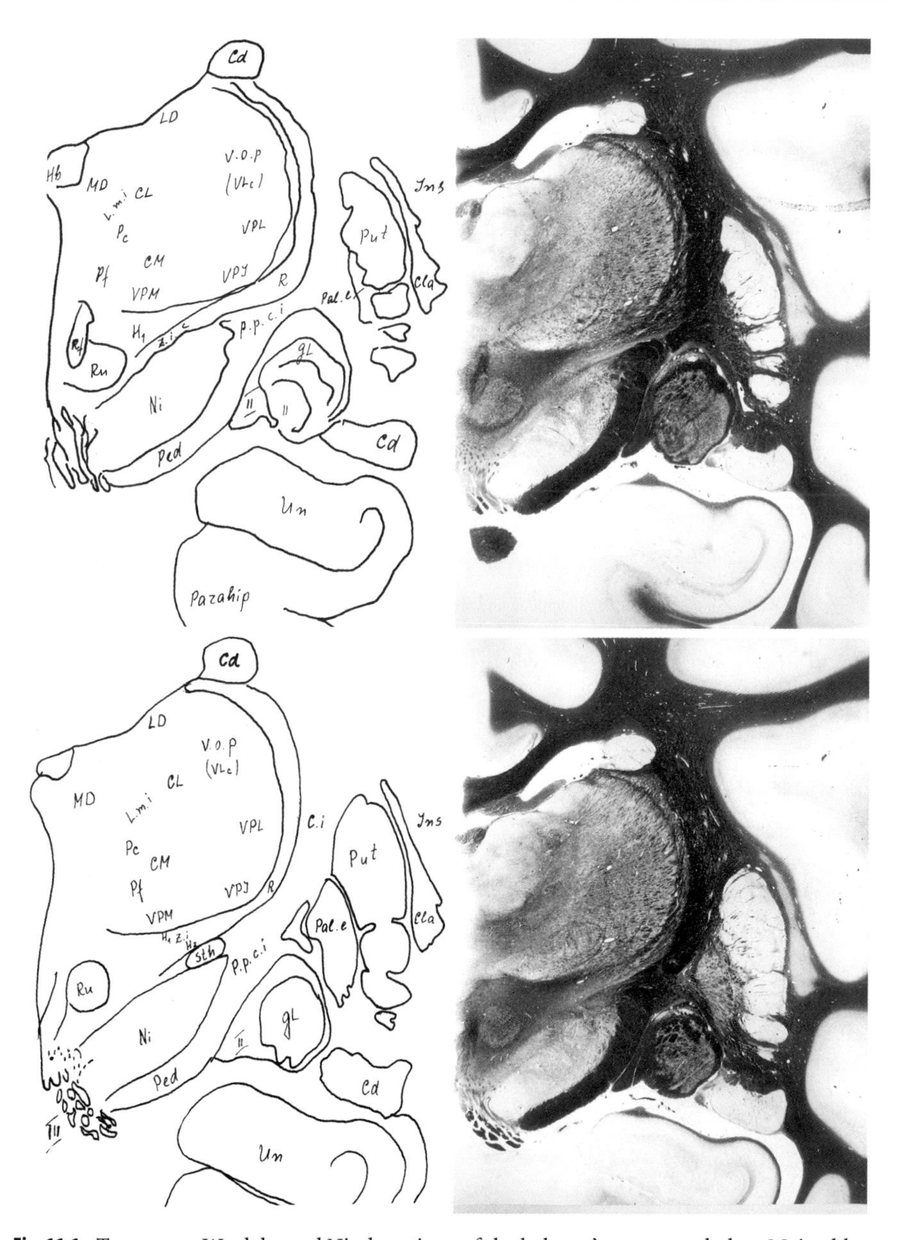

Fig. 11.1 Transverse Woelcke and Nissl sections of the baboon's mesencephalon. Main abbreviations (also see abbreviations list): *cd*, nucleus caudatus; *ci*, capsula interna; *cla*, claustrum; *H*, *H1*, *H2*, fields of Forel; *Ni*, *Nic*, *Nir*, substantia nigra, pars compacta, pars reticulata; *pale*, *pali*, globus pallidus externus and internus, respectively; *ppci*, pars peduncularis of the capsula interna; *put*, putamen; *Ru*, nucleus ruber; *STH*, nucleus subthalamicus; *TcTT(R)*, tractus corticotegmentothalamicus (Rinviki); *VPM*, *VPI* and *VPL*, nucleus ventralis posterior, intermediate and lateralis thalami; *Zi*, zona incerta; *II*, optic tract

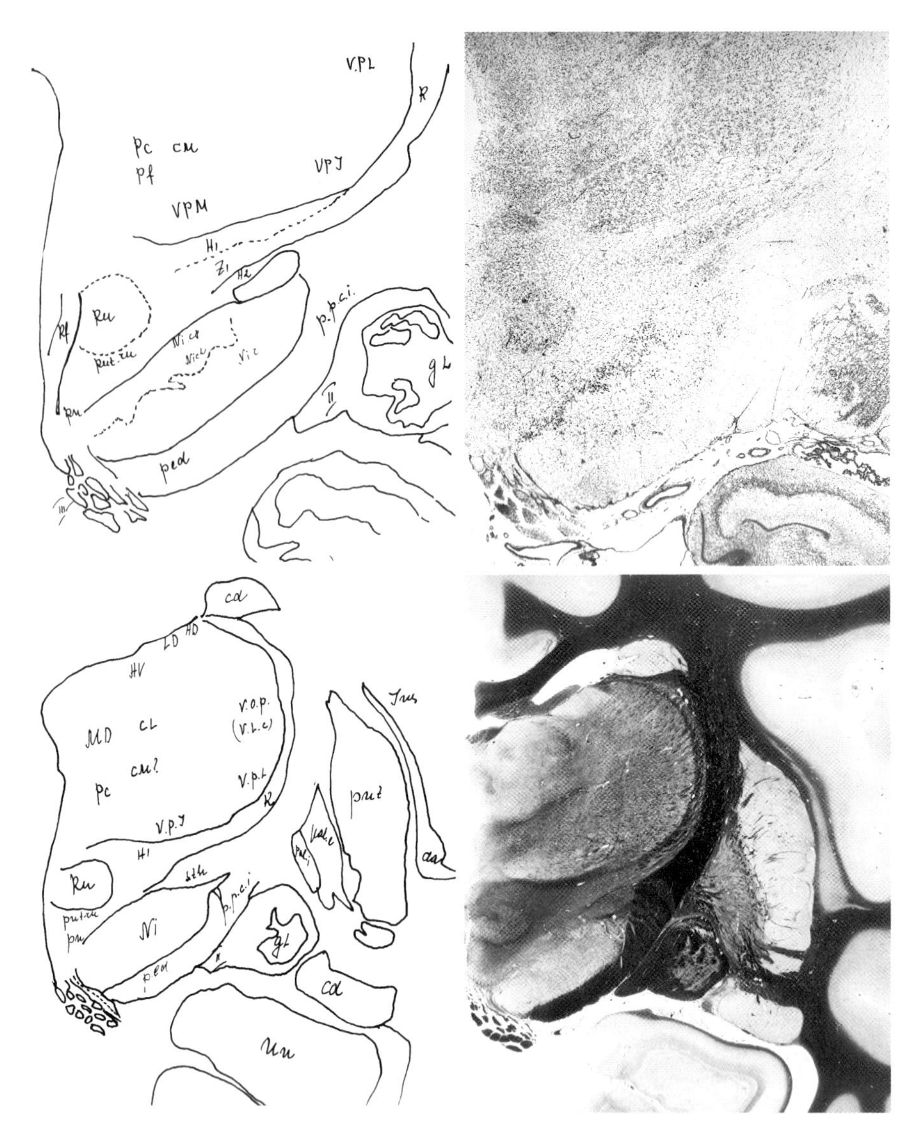

Fig. 11.2 (continued)

The comb system of Edinger has nearly reached maximum extent. The head of the zona incerta has the dimensions of its cauda and is separated from the STN by field H2.

The internal capsule contains subdivisions of the comb system: most laterally the fields A of Sano (1910), more medially the rest of the comb system. In the lateral fields between the rostrolateral pole of the pars reticulata and the caudal medial

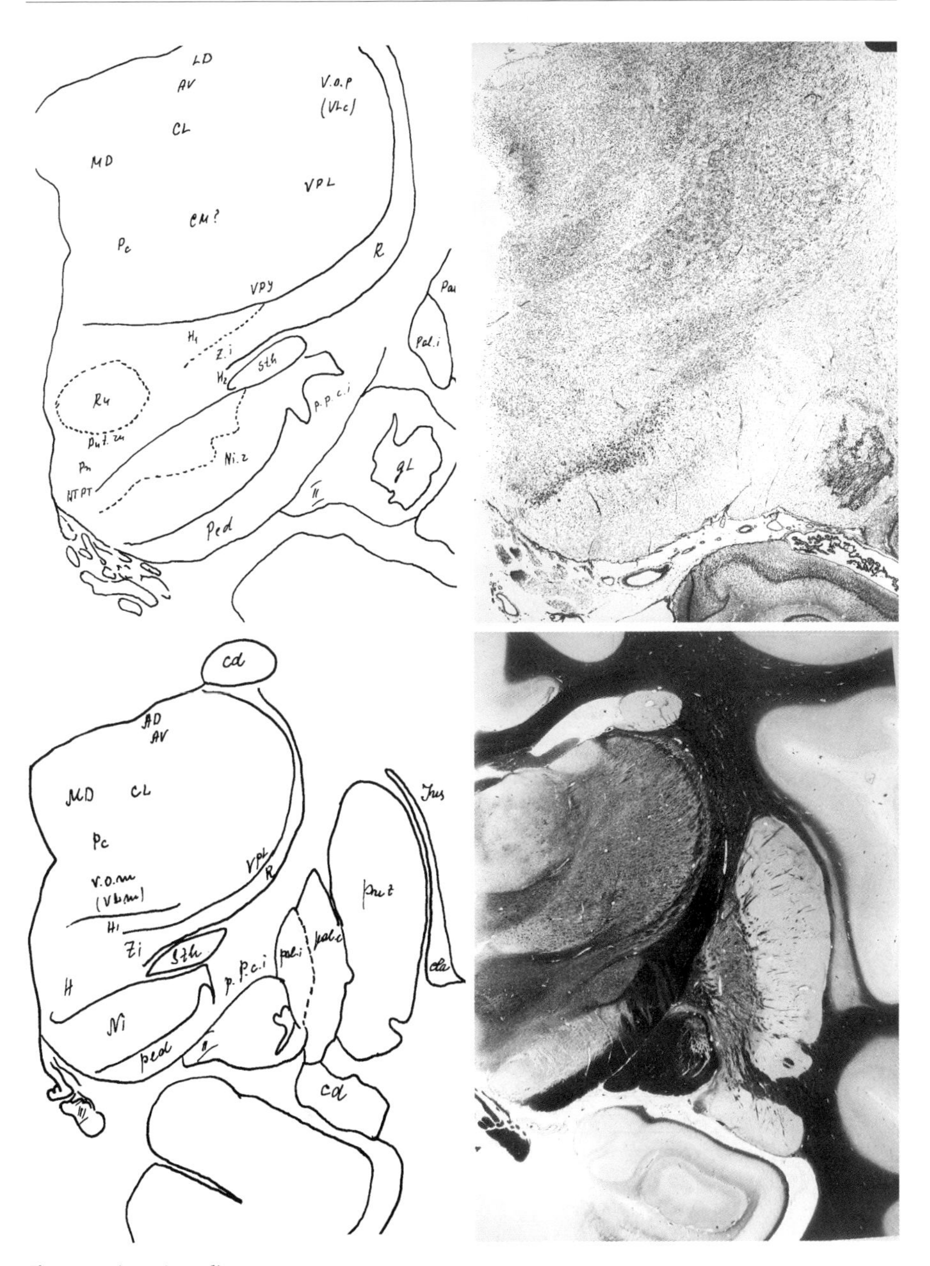

Fig. 11.3 (continued)

pole of the internal pallidum isolated neurons are occasionally present. This is the field A of Sano (1910), who was the first to speculate that these interstitial neurons represent a feeble cellular bridge between the substantia nigra and the pallidum. Later, Riese (1924) noticed in cetaceans that the pars reticulata is strongly extended

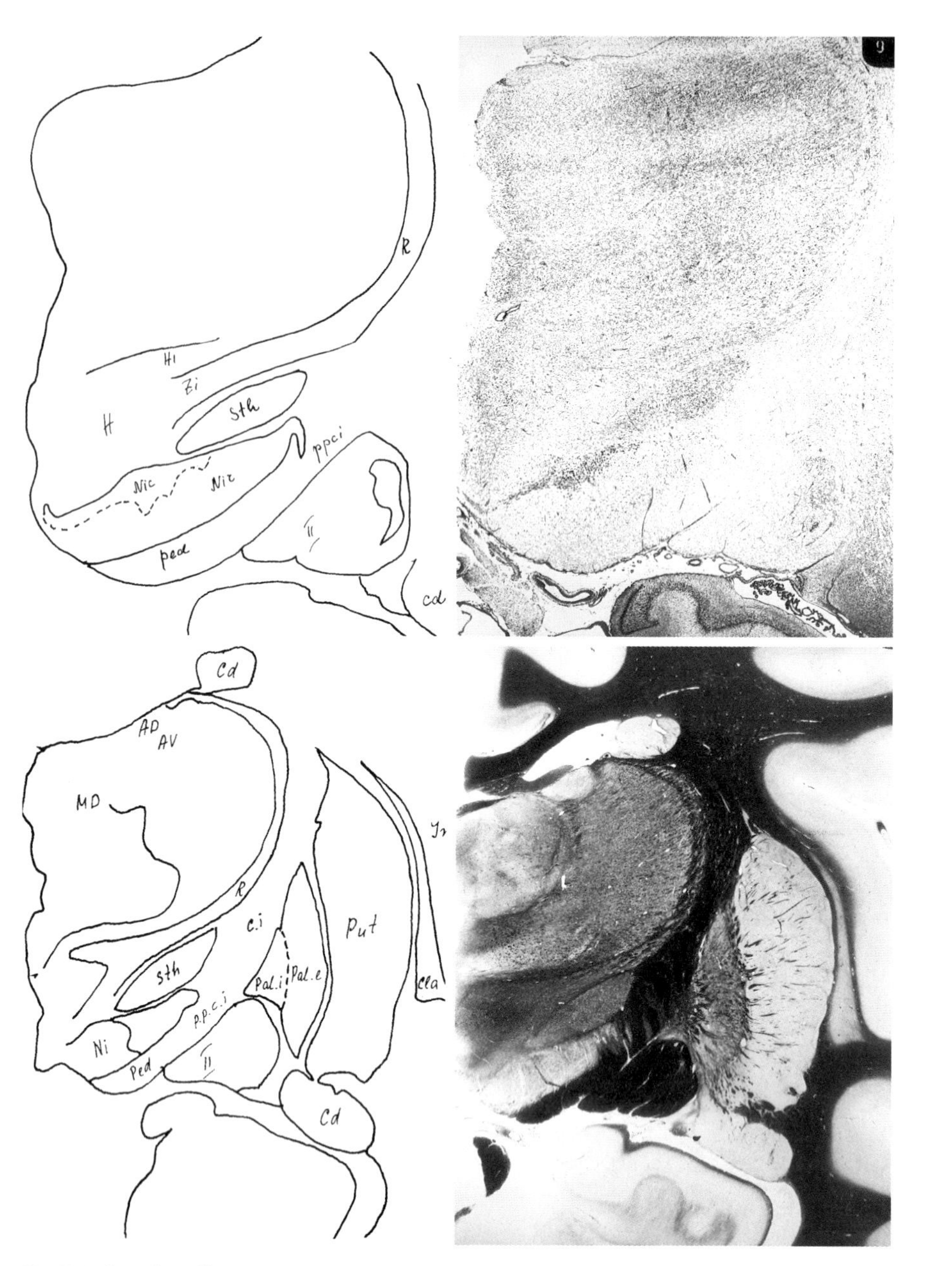

Fig. 11.4 (continued)

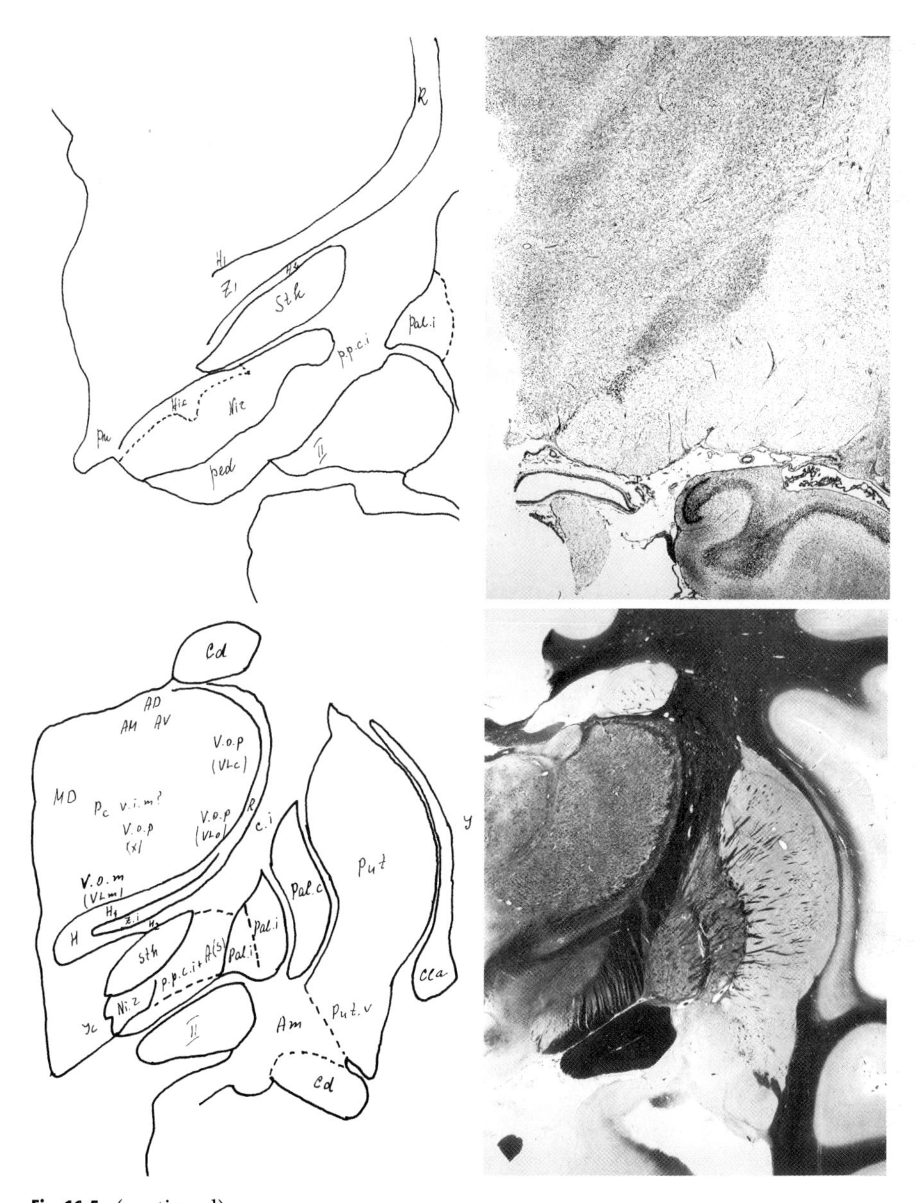

Fig. 11.5 (continued)

rostrally and especially rostrolaterally, so that its grey strands intermingle with the comb system and directly continue within the pallidum.

4.4
The Human Subthalamic Nucleus: Topography

Two sections (Fig. 12) that are on the oblique coronal plane and parallel to the course of the optic tract (left Nissl, right myelin stain; Voogd and Feirabend 1981) show the sagittal topography of the human STN. In these sections the relation with the substantia nigra cannot be discerned. The human STN is sagittally sectioned parallel to the optic tract. The STN lies in a hollow of the cerebral peduncle and is covered by a thin layer of fibres, the field H2 of Forel, while separated from the cerebral peduncle by a small fibre layer. At the medial side the fornix and tractus mamillo-thalamicus can be discerned. The zona incerta covers with head and tail the STN. The hypothalamus is between the medial part of the peduncle, ventricle and the optic tract. In both sections the comb system of Edinger can be discerned.

The human Nissl sections and myelin-stained sections (30–40, going from rostral to caudal) correspond to Mai and colleagues' *Atlas of the Human Brain* (Mai et al. 1997). In the atlas the corresponding section numbers are explained.

4.4.1
Nissl Sections

In a topographic sense the pallido-hypothalamic nucleus (see Mai et al. 1997) borders the caudal side of the STN (sections 30 and 31, Fig. 12B). Its borders are difficult to delineate, which means that in man the STN at its caudal side seems continuous with a part of the lateral hypothalamus. Groenewegen and Berendse (1990) noted projections from the ventral pallidum towards the STN and the lateral hypothalamus (see Sect. 5.2.4, this volume). The relation between STN and the lateral hypothalamus can also be clearly discerned in the oblique coronal sections of Fig. 12A. These projections attracted attention, but no available arguments in human degeneration studies have suggested that these connections are present in man, although they are established in rat.

The STN over-roofs the substantia nigra pars compacta and pars reticulata with its caudal extension (section 32). Nearly directly (section 33), the supramammillary commissure (see Mai et al. 1997) pushes the STN laterally (see also the myelin stained section 33 Fig. 12C), separating the STN at its caudal side from the substantia nigra (section 33, Fig. 12B). Directly rostrally, after this commissure, the former relations are re-established.

At the rostral side of the STN (sections 38–40, Fig. 12B) the rest of the STN seems still present between the capsule of the red nucleus, the parabrachial pigmented nucleus and the substantia nigra (section 38, Fig. 12B). In sections 39 and 40 (Fig. 12B) the parabrachial pigmented nucleus seems to over-roof the whole

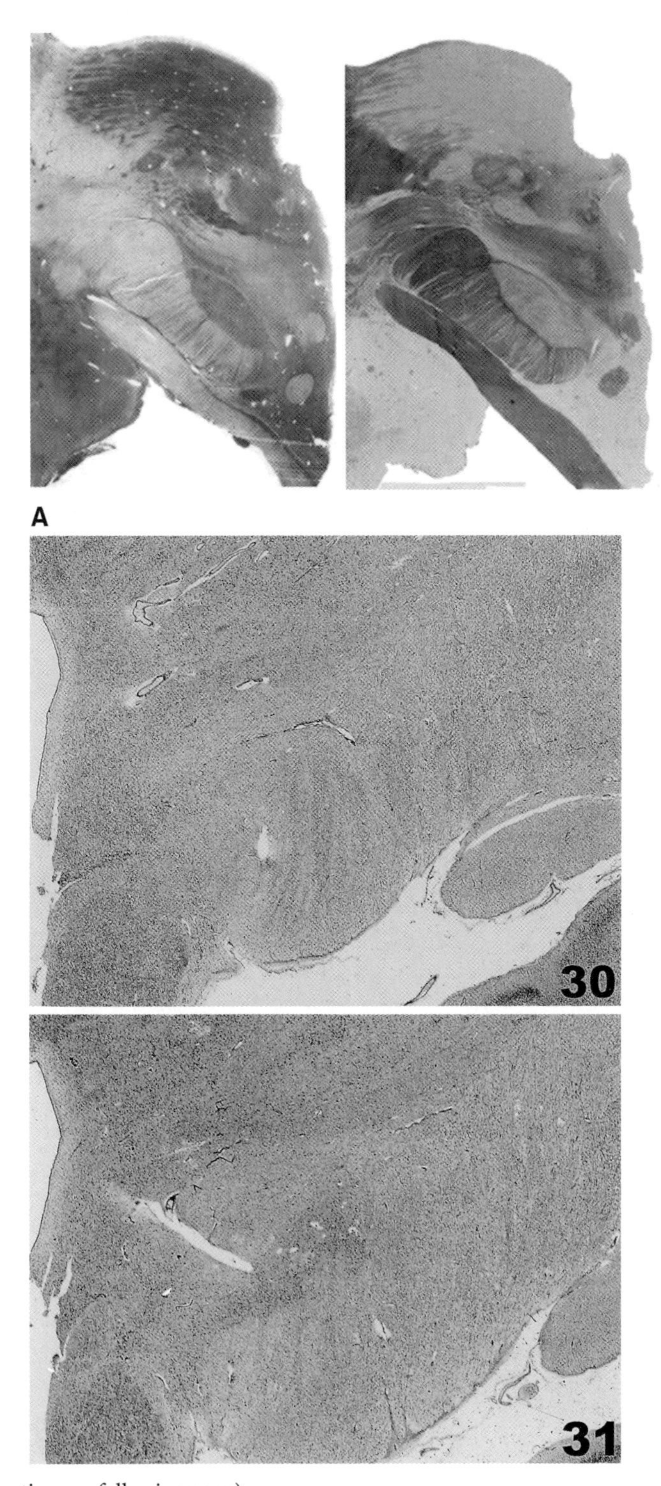

Fig. 12 (Caption see following page)

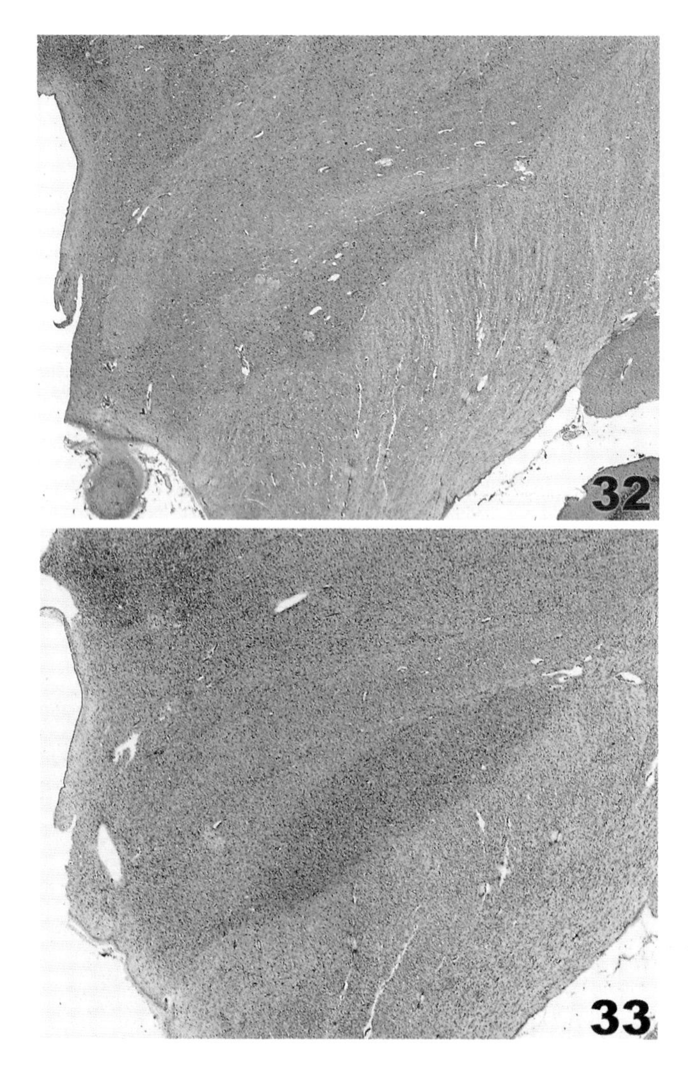

Fig. 12 A Coronal sections through the human STN. *Left*, Nissl stain; *right side*, myelin staining. B Transverse serial sections of the human STN (Nissl stain) starting at the caudal/substantia nigra side. C Sections, stained for myelin, of the human STN. The panels match with sections numbers of the *Atlas of the Human Brain* (Mai et al. 1997)

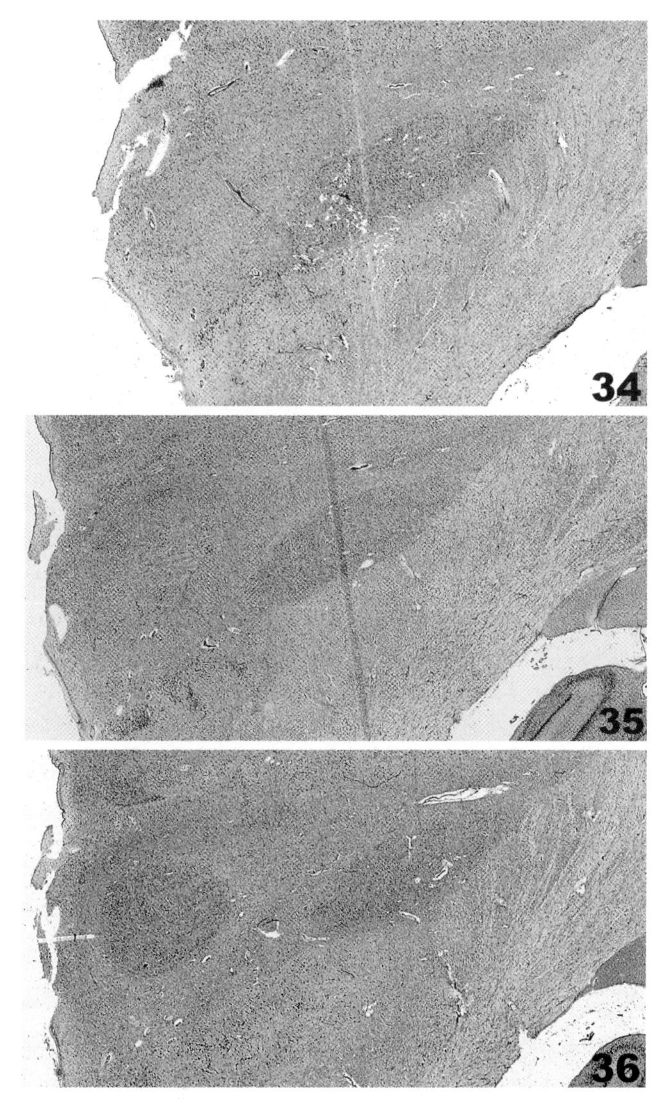

Fig. 12 (continued)

substantia nigra. Over its whole rostral extent the STN is covered by the zona incerta, which retracts due to the origination of the parabrachial pigmented nucleus (compare to the *Atlas of the Human Brain*; Mai et al. 1997).

While in the rat the origin of the STN clearly could be seen arising in between the peduncular bundles, this is obscured by the human pallido-hypothalamic nucleus, which also has its neurons between peduncular bundles. Nevertheless section 38 in Fig. 12B also shows the pars caudalis of the STN to be present in between

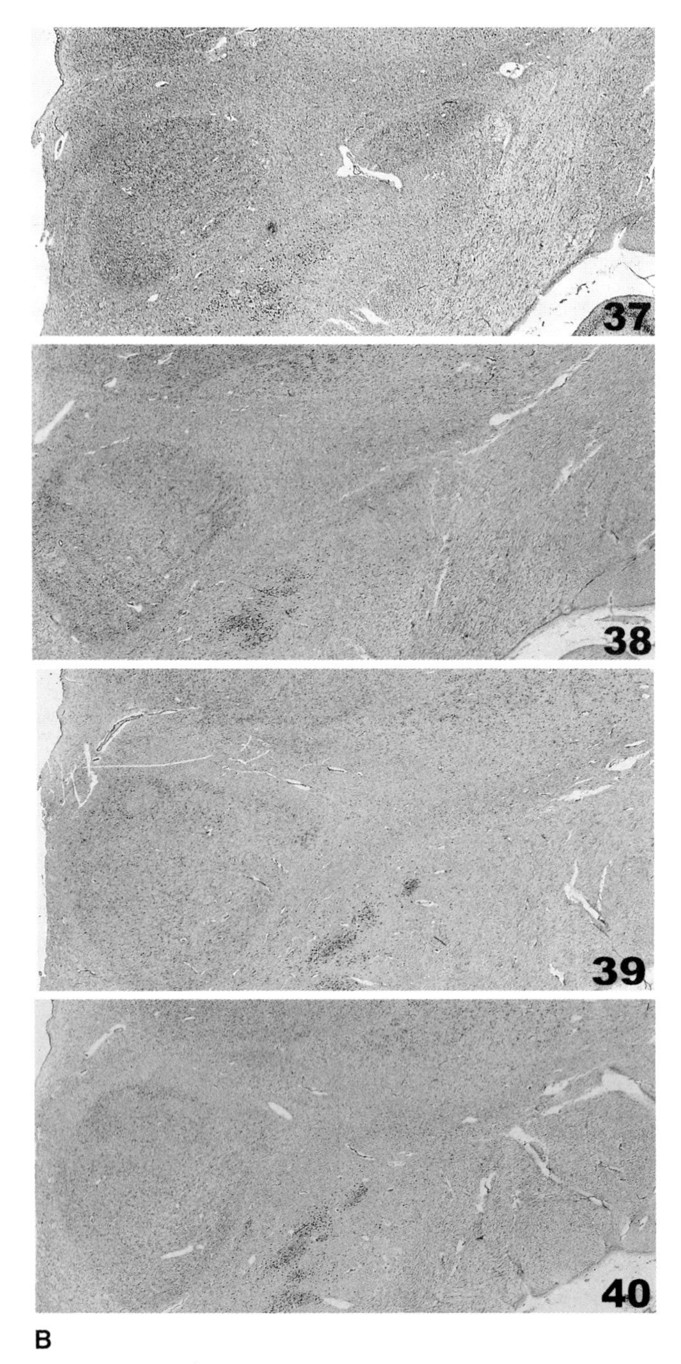

Fig. 12 (continued)

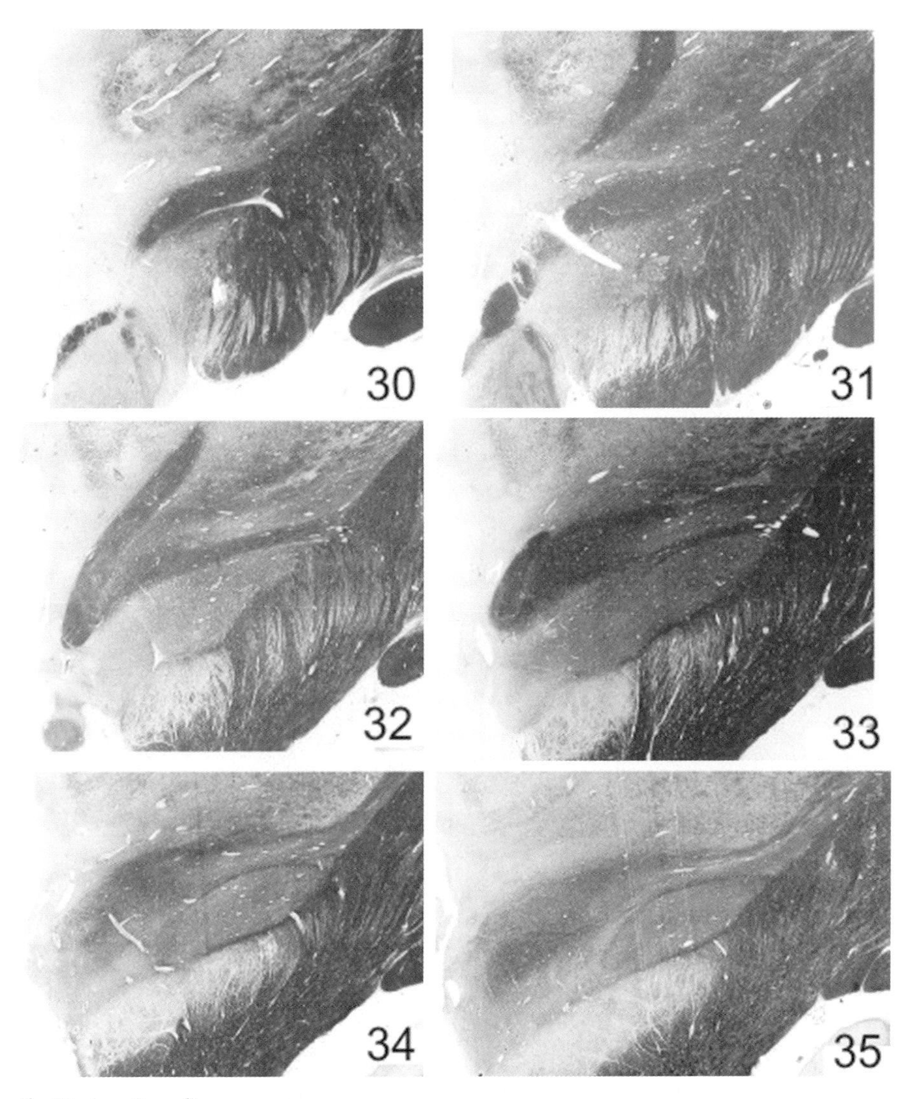

Fig. 12 (continued)

peduncular bundles. The human STN clearly has a lens-shaped form. In the middle part of section 33 an area with a lower concentration of neurons can be detected, which proceeds laterally in section 34, ending in section 35.

Due to the caudal position of the human STN, a relation between the substantia nigra and the STN starts no earlier than in section 34, indicating that the gradual transition between both nuclei as present in rat, cat and baboon is absent in humans. The caudal part of the STN can be recognized independently from the substantia nigra. This relation is even more important because in T_2-weighted

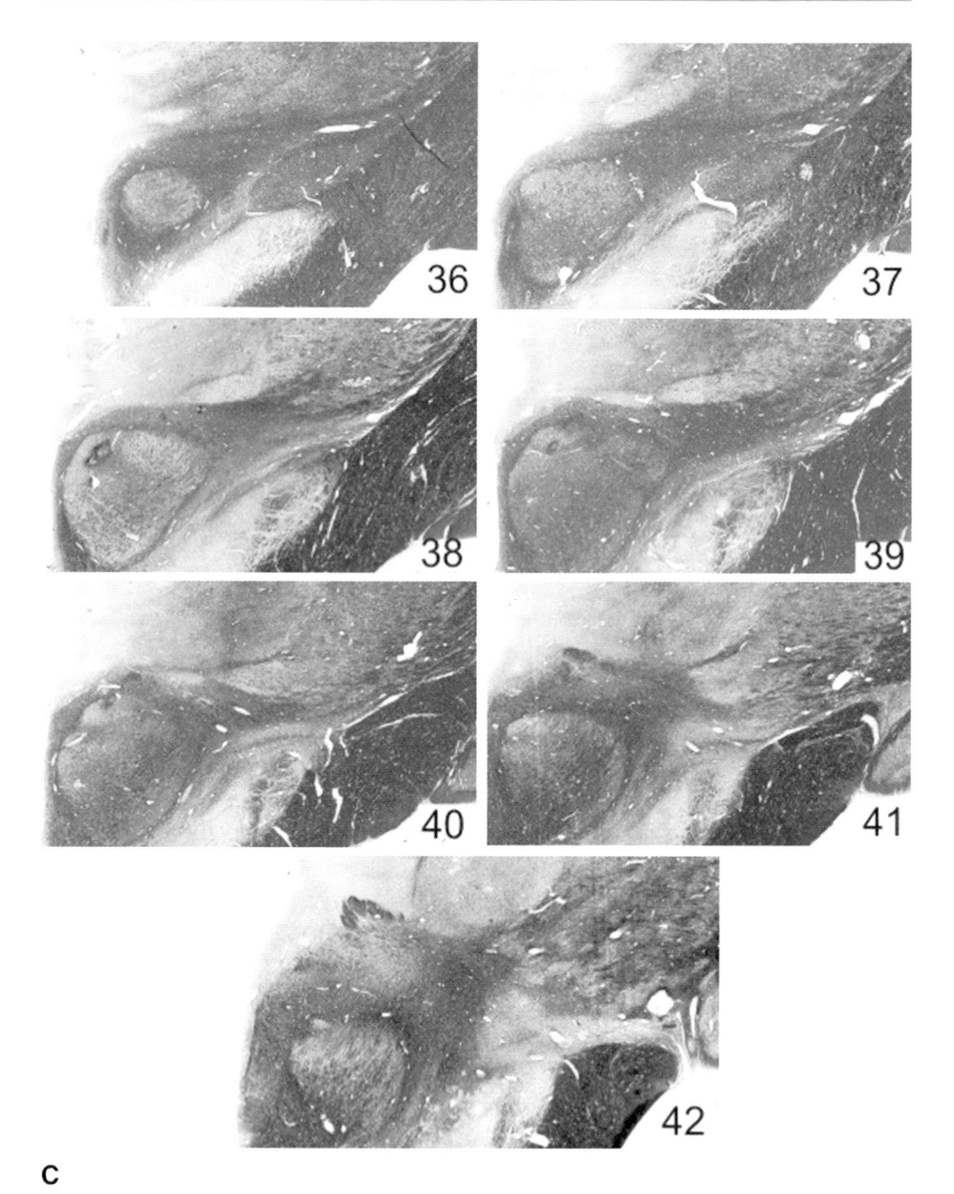

Fig. 12 (continued)

images the substantia nigra, due to its high iron content, can be recognized easily, while the anterior medial part of the STN is more difficult to perceive. The contact between substantia nigra and STN stays present from sections 34 to 37 (Fig. 12B). In section 38 the last part of the STN disappears next to the rostrolateral part of the substantia nigra. Whether or not some cells are still present in section 39 is uncertain, if solely based on Nissl stainings; see also picture 39 of Mai's atlas (Mai et al. 1997). At level 40 the peripeduncular area distinctly protrudes between the zona incerta and the substantia nigra. Therefore the STN has ended.

A third point is the relation towards the red nucleus. The STN has already been developed before the red nucleus appears (section 32–34, Fig. 12B). Since the red nucleus is simple to find in T_2-weighted images, care should be taken not to underestimate the extent of caudal STN.

4.4.2 Myelin-Stained Sections

The myelin-stained sections show a cap of myelinated fibres at the lateral half/two-thirds of the nucleus (see sections 33–35, Fig. 12C; for explanation of the involved systems see Sect. 5.2, this volume). The dorsal side is made up by the slimming field H_2, while at the ventral side the rim reaches from the cerebral peduncle over the substantia nigra. The ventral rim clearly borders the pars reticulata of the substantia nigra.

Myelinated fibres can be followed from this cap intruding into the lateral edge of the STN. The fibres penetrate along the long axis of the nucleus. Not earlier than in section 36, field H_2 comes to an end and is replaced by fibres of the zona incerta. Therefore, the coverage of the STN by myelinated fibres is not only by field H2, but at its lateral rostral one-third by zona incerta fibres. It is here (sections 37 and 38, Fig. 12C) that bundles of the cerebral peduncle seem to separate from the body of the peduncle due to the interposing of the substantia nigra. It looks like these bundles traverse the substantia nigra to keep up the ventral rostral rim at the STN.

Special attention is given here to the Kammsystem of Edinger (see Sect. 5.2, this volume). In the transitional area of pedunculus cerebri towards the capsula interna, that is just dorsal to the optic tract, the comb system starts (section 36). Going ventral caudalwards, it stretches out into the cerebral peduncle to end in section 30. Over its whole rostral half the comb system stays localized ventral of the STN with its fibres pointing perpendicular to the long axis of the STN. The rostral side, which increases towards the substantia nigra, pushes the comb system backwards, but an intimate relation with the STN stays present.

4.5 Ageing of the Human STN

The human STN was studied by den Dunnen and Staal (2005) in 12 post-mortem brains of patients who died of non-neurological diseases. The group comprised five females and seven males. Using the anterior commissure–posterior commissure (AC–PC) line as the reference—and independent of gender, side of the brain and length of the AC–PC line—displacement of the centres and borders of the STN were found. The superior–inferior direction of the STN shows that the STN moves cranially and becomes smaller in this direction during ageing, if the extreme ages (29 years versus 84 years) are compared. In the medial lateral direction the STN moves further away from the midline with increasing age. Moreover, the diameter in this direction increases. In the anterior–posterior direction the diameter decreases with increasing age. The diameter decrease in the superior inferior direction is nearly 2.2 mm or over 30% of the youngest diameter. The medial–lateral diameter increase is over 25%, while the reduction in the anterior posterior direction is also over 25%. Changes of the centre positions of the STN are: superior–inferior over 50%, medial–lateral 5% and anterior–posterior 70%. However, since the distance shift in the anterior–posterior direction is in absolute terms small (0.5 mm), while superior-inferior it is 3.9 mm and medial lateral 2.6 mm the stereotactic implantation of electrodes is influenced most by the superior–inferior and medial–lateral directions. During ageing the nucleus changes from spindle-shaped to discus-shaped.

Going from rat to the baboon it can be noticed that the STN is placed more over the substantia nigra in the primate (including man). Within the rat STN an architectonic subdivision can be made. A medial, intermediate or central and a lateral part are discerned. The rat STN cells disappear between the peduncular fibres, and this part is called pars dissipata. In the baboon a tripartition of the STN can be found in Nissl sections too, based on the concentration of cells. This holds for man as well. The start of the human STN is difficult to discern due to the localization of the pallido-hypothalamic nucleus, while a pars dissipata as in rat is difficult to discern. The human STN reaches far more caudally compared to the other species. Moreover, a clear lateral cap of myelinated fibres is present around the human STN. The intimate relation between the rostral part of the comb system and the STN is evident.

5 Connections of the Subthalamic Nucleus

5.1 Overview of the Mature Connections

The STN was believed to exert an inhibitory, probably GABA- or glycine-mediated, effect on its target nuclei, and this common belief persisted for years (Yoshida 1974; Brodal 1981; see also Mehler 1981), which was one of the major reasons to overlook the involvement of the STN in the parkinsonian pathophysiology. It is now firmly established that the STN projection neurons are glutamatergic, excitatory (Hammond et al. 1983a, b; Kitai and Kita 1987; Smith and Parent 1988; Albin et al. 1989a, b; Robledo and Feger 1990; Brotchie and Crossman 1991; Rinvik and Ottersen 1993; Feger et al. 1997), and heavily emitted by widely branching axons: the substantia nigra (SN) and the internal pallidal segment (GPi), followed by the external pallidal segment (GPe) and the pedunculopontine tegmental nucleus (PPN). Leucine-labelled fibres of the STN follow in their projections the laminar organization of the substantia nigra's pars reticulata (Tokuno et al. 1990). Some STN axons reach the neostriatum (Carpenter and Strominger 1967; Kanazawa et al. 1976; Nauta and Cole 1978; van der Kooy and Hattori 1980; Carpenter et al. 1981; Hammond et al. 1983a, b; Moon Edley and Graybiel 1983; Kitai and Kita 1987; Parent and Smith 1987; Takada et al. 1988; Groenewegen and Berendse 1990; Smith et al. 1990b; Shink et al. 1996; Sato et al. 2000b). There are also unconfirmed reports that the STN innervates directly the cerebral cortex (Jackson and Crossman 1981), and even the spinal cord (Takada et al. 1987).

The most prominent afferent connections of the STN arise in the GPe. The pallido-subthalamic GABAergic boutons almost completely cover the perikarya of the STN projection neurons and their proximal dendrites (Fig. 13; Nauta and Mehler 1966; Carpenter et al. 1968, 1981; Nauta 1979; Romansky et al. 1980a, b; Mehler 1981; Usunoff et al. 1982b; Romansky and Usunoff 1985, 1987; Smith et al. 1990a, 1998; Bell et al. 1995; Shink et al. 1996; Sato et al. 2000a). The STN is also innervated by the cerebral cortex, and the glutamatergic corticosubthalamic axons terminate mainly on the distal dendritic portions of the projection cells, and on the vesicle-containing dendrites of the interneurons (Künzle 1978; Romansky et al. 1979; Kitai and Deniau 1981; Romansky and Usunoff 1987; Canteras et al. 1988). Area 4 projects somatotopically to the STN and to areas 6 and 8 topologically (Hartmann-Von Monakow et al. 1978). A substantial, bilateral cholinergic/glutamatergic projection

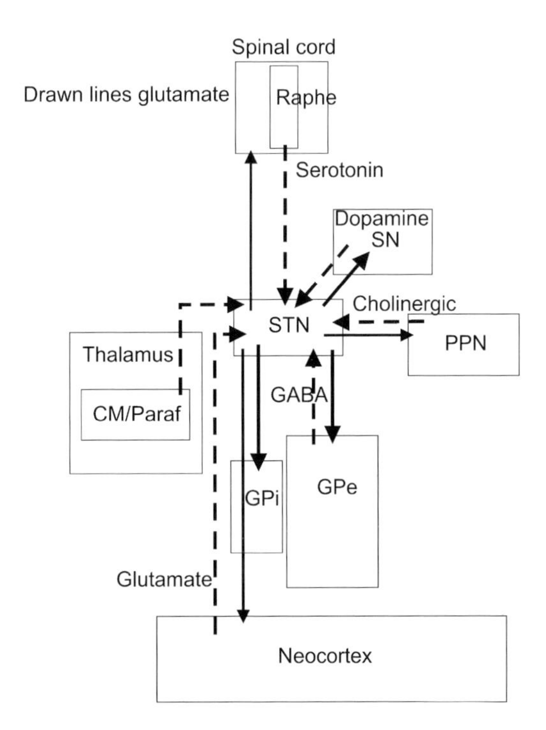

Fig. 13 Systemic drawing summarizing the overview of the STN afferent and efferent connections

arises in the PPN, and its endings perform both axosomatic and axodendritic synaptic contacts (Hammond et al. 1983a; Jackson and Crossman 1983; Moon Edley and Graybiel 1983; Romansky and Usunoff 1983, 1987; Lee et al. 1988; Lavoie and Parent 1994a, b; Bevan and Bolam 1995; Takakusaki et al. 1996; Smith et al. 1998). The thalamic centromedian-parafascicular complex also innervates the STN (Sugimoto et al. 1983; Sadikot et al. 1992). Finally, serotoninergic fibres from the raphe nuclei terminate profusely within the STN (Mori et al. 1985; Lavoie and Parent 1990; Leger et al. 2001).

The nigrosubthalamic connection was not demonstrated by means of the Nauta silver impregnation techniques, and even the more effective Fink–Heimer technique and its modifications provided only a vague evidence for the existence of such a connection (for a review see Usunoff et al. 1976). Only the modern axonal transport techniques combined with transmitter immunocytochemistry offered firm evidence for existence of this pathway (Brown et al. 1979; Meibach and Katzmann 1979; Rinvik et al. 1979; Gerfen et al. 1982; Björklund and Lindvall 1984; Lavoie et al. 1989; Overton et al. 1995; Hassani et al. 1997; Cossette et al. 1999; Gauthier et al. 1999; Hedreen 1999; Francois et al. 2000; Ichinohe et al. 2000; Prensa and Parent 2001). Although appreciated differently (from scant to strong), most of the studies agree that the nigrosubthalamic connection arises from the dopaminergic (DAergic) neurons of the SN pars compacta (SNc), and this conclusion is supported by the report on DA receptors in the STN (Flores et al. 1999). There is a recent suggestion that DA exerts a direct excitatory influence on STN neurons via the activation of D2-like receptors (Zhu et al. 2002). In addition, Ichinohe et al. (2000) have described a moderate projection from the parvalbumin immunoreactive, presumably GABAergic neurons (see Parent et al. 1995; Gerfen and Wilson 1996) of the SN pars reticulata (SNr) to the STN. The nigrosubthalamic connection has always been described as ipsilateral, even by Gerfen et al. (1982), who eagerly examined the crossed connections of SN. We currently report that a substantial, bilateral projection of the SN terminates in the STN of the rat (see Sect. 6, this volume).

5.2 Afferent and Efferent Human Connections

5.2.1 Cortical Connections

What do we know of the motor-cortical connections with the STN in man? Luys was the first to propose the human corticosubthalamic connections (see Parent et al. 2002). In *Folia Psychiatrica Neurologica et Neurochirurgica Neerlandica*, Stenvers (1953) made a clinical anatomo-physio-pathological contribution to human pyramidal and extrapyramidal disorders. The article contains an overview of the connections involved in choreoathetosis, tremor hypertonus and ballism, based on human material. His corticofugal fibre figure distinctly shows the presence of a cortical contribution to the nucleus subthalamicus and it is described in the text. Thus, around 1955 acceptance of the presence of these connections prevailed.

Unlike animal experiments, in human material the extent of the lesion cannot be manipulated. Therefore large, but sometimes even small bleedings or lesions are studied.

The fronto-corticosubthalamic pathway was suggested on the basis of examination of human material obtained from patients after prefrontal lobotomy for surgical treatment of schizophrenia (Meyer 1949). However, experimental Marchi studies (Levin 1936; Verhaart and Kennard 1940; Mettler 1947) failed to demonstrate cortical fibres terminating in the STN.

Ten years later experimental animal studies showed the reverse. The most extensive description in experimental animal studies is given for the rat by Knook (1965).

After cortical ablations, a large contingent of fibres is seen in the internal capsule and pedunculus cerebri (Fig. 14A). At subthalamic levels many fibres leave the capsule. Three groups leave above the STN: (1) bundles entering the thalamus via the nucleus reticularis; (2) a series of bundles penetrate the nucleus reticularis thalami that continues in the lamina medullaris ventralis thalami, intermingling with the fibres of the lemniscus medialis; (3) a series of bundles that pass directly from the internal capsule. These fibres pass through the zona incerta and partially end in it. (4) The most ventral

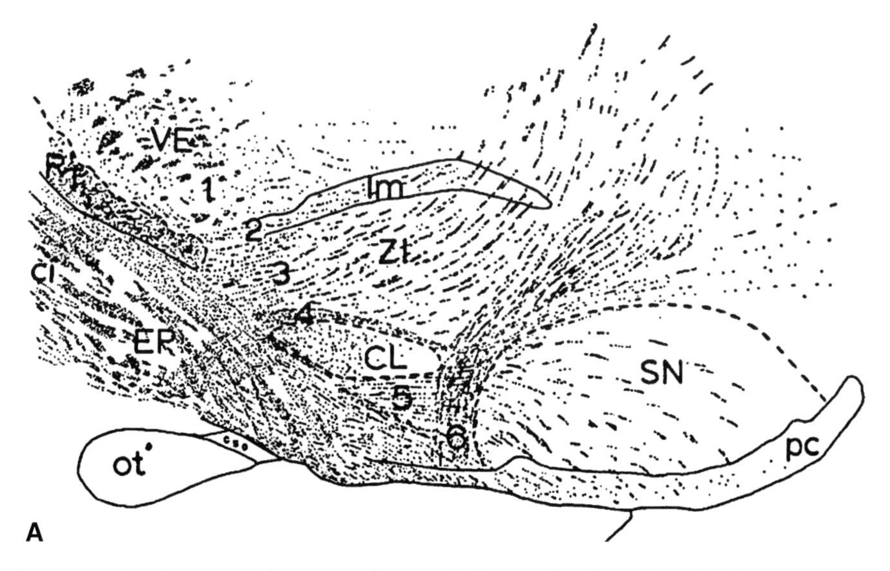

Fig. 14 A Distribution of degenerated cortical fibres related to the subthalamic area. *Ci*, intern capsule; *CL*, corpus Luyisii, subthalamic nucleus; *ER*, entopeduncular nucleus; *lm*, medial lemniscus; *pc*, pedunculus cerebri; *RT*, nucleus reticularis thalami; *ot*, optic tract; *SN*, substantia nigra; *VE*, nuclei ventralis thalami. **B** Fink-Heimer degeneration after rat cortical lesions. *Upper row*, cortical lesions, *Middle row, left*, degeneration around the subthalamic nucleus (compare to Fig. 12A); *right*, corticofugal degeneration passing through the substantia nigra. *Lower row, left*, returning cortical degeneration into the cerebral peduncle; *right*, cortical degeneration in the pars compacta (see Malinov et al. 1984)

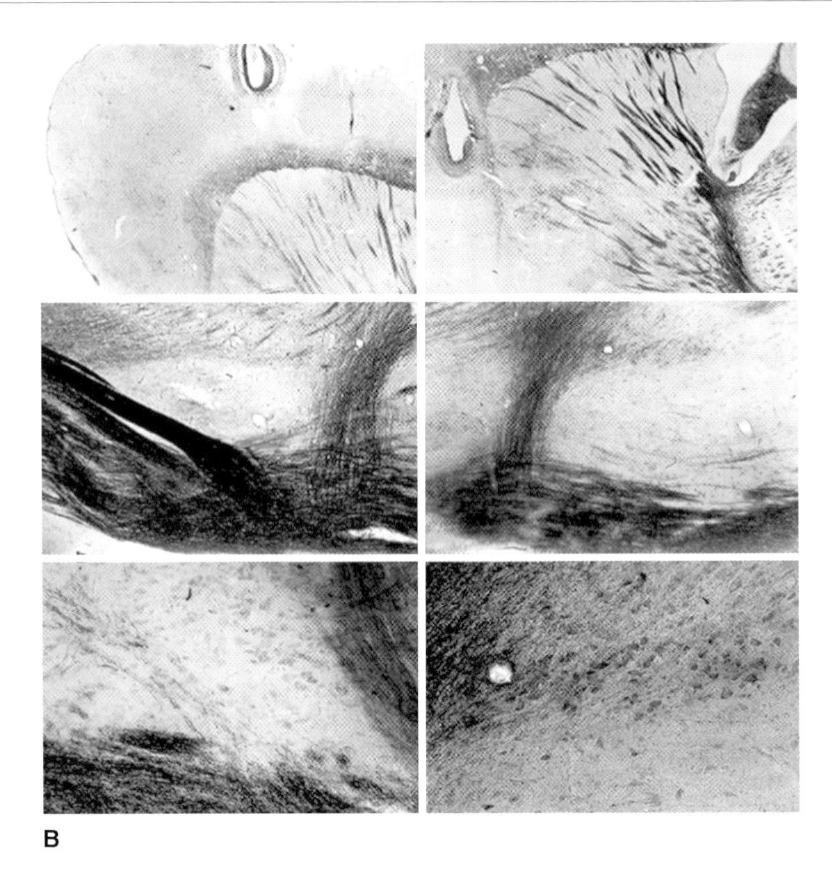

Fig. 14 (continued)

fibres enter the "subthalamus by way of the internal capsule, course caudally along the dorsal surface of the corpus Luysii, to which many fibres are distributed" and pass further to the substantia nigra. (5) A fifth component leaves the peduncle ventral to the STN and courses caudally through the substantia nigra. "A small part of its fibres, however, follows the dorsal border of the nigra and joins the component (4)". Nevertheless, (6) the most substantial component of fibre bundles leaves the peduncle to find their way in between the STN and the substantia nigra and fan out over the entire mesencephalic reticular formation (Fig. 14A and B). The fibres that pass through the STN and substantia nigra return into the cerebral peduncle (Fig. 14B).

Dejerine (1901) found that cortical fibres can be traced into the STN in humans, but Levin (1936) identified these as corticorubral fibres based on his study in monkeys. Kodama (1926) denied the existence of corticosubthalamic fibres in human; however, it is possible that the medial part of the STN can receive some ("spärlicher Anzahl") cortical fibres. In 1929 Wilson described a case of hemiballism in which the *contralateral* postcentral gyrus was lesioned solely. "Although this report is unique in literature, it must be retained for at least tentative consideration in view of Wilson's reputed thoroughness and his penetrating analytic and self-critical capacities" (Meyers 1968). Meyer (1949) found in human brains no projection into the STN as long as the leucotomy lesion did not reach the white matter of the prefrontal areas. If these lesions reach into the white matter, degeneration was indeed present in the STN. This would underlie that the dysgranular and agranular cortex of the frontal terminals was responsible for the STN degeneration. Mettler (1945) had earlier found no changes in the STN if the cortex or most of the striatum were removed in monkey and baboon, but they found that pallidal lesions do effect such changes. Moreover, Gebbink (1967) could not confirm the corticosubthalamic connection in humans with the Nauta method. Hence, the old literature up to the 1970 is inconclusive on the corticosubthalamic connection in man. Moreover, Parent and Hazrati (1995a, b) in their excellent description on the place of the STN in the basal circuitry omit any reference to human corticosubthalamic connections. In fact Strafella et al. (2004) indicate that the role of the cerebral cortex in regulating the activity of the STN "is not known in humans". Transcranial magnetic stimulation of the motor cortex changes the neuronal activity in the STN. Stimulation of the motor cortex in Parkinson's patients undergoing implantation for deep brain stimulation shows an induced excitation in 75% of the neurons investigated. The STN neurons responsive to motor hand cortex area stimulation were localized mainly in the lateral and dorsal areas of the STN (Fig. 15). Neurons unresponsive to this stimulation were located more medially.

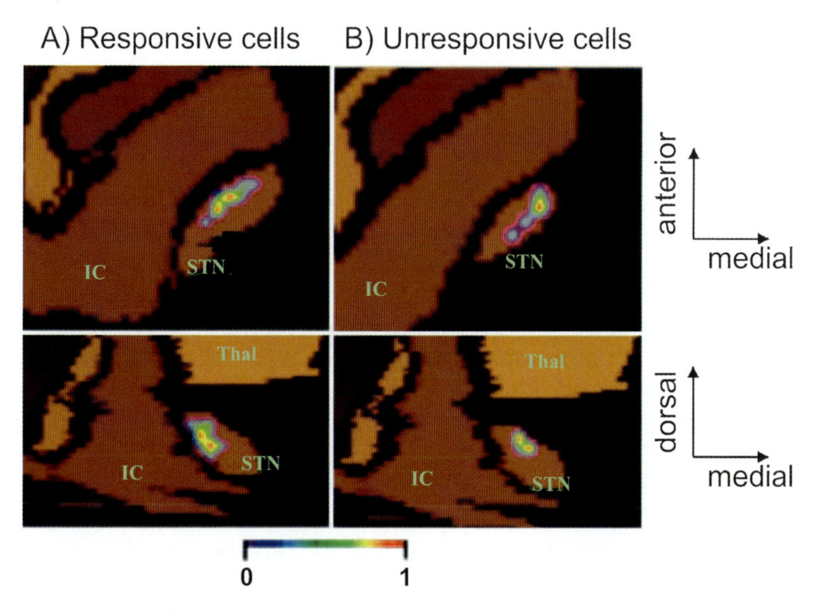

Fig. 15 Horizontal and coronal sections of the MR image atlas. Lateral localization of responsive cells (**A**), and unresponsive cells (**B**). (Courtesy Strafella et al. 2004)

Other experimental animal studies before 1980, carried out with the Nauta technique (Auer 1956; De Vito and Smith 1964; Petras 1965) and with autoradiography (Künzle 1976; Künzle and Akert 1977) or with HRP (Rinvik et al. 1979), confirmed the existence of the corticosubthalamic projection in a series of species.

Comparative experimental animal research on the corticosubthalamic connections shows that in Kalong (*Pteropus giganteus*) and Capybara (*Hydrochoerus hydrochaeris*) these fibres end homolaterally in the STN. In the sub-primate Tupaia (common tree shrew) an ipsi- and contralateral ending of this system was found (Broere 1971). The opossum (*Didelphis* spec.) contains cortical endings in the STN (Cabana and Martin 1986). In the goat pericentral cortical ablations show degeneration in the ipsilateral rostral half of the STN (Haartsen 1962).

Künzle (1976) in his study on area 4 of the precentral gyrus in Macaca showed using ^{3}H leucine and ^{3}H proline that cortical projections arrived in the STN. This proved that terminal fields originating from cortical neurons were antegradely labelled into the STN. Moreover, injections into face, arm, and leg areas of cortical area 4 showed a somatotopy in the lateral part of the STN (Künzle 1976). Face projections are localized more ventro-laterally and arm and leg more dorso-laterally, although serious overlap was noticed. Area 4 indeed projects somatotopically to the STN according to Hartmann-Von Monakow et al. (1978), confirming the earlier results. These somatotopic results in monkeys were confirmed by Carpenter et al. (1981), also in monkeys.

A second wave of publications on the corticosubthalamic connections appeared around the 1990s. Most of them concerned the rat and a few monkeys and are electrophysiological or retrograde tracing-driven (Rouzaire-Dubois and Scarnati 1985; Canteras et al. 1990; Nambu and Llinas 1994; Fujimoto and Kita 1993; Mouroux et al. 1995; Maurice et al. 1999). All these articles confirmed what was already described in the autoradiographic and Nauta degeneration results. In general the motor, sensory and frontal lobe cortex project to the STN in rats, cats, and monkeys, the strongest being the motor cortex (Afsharpour 1985a, b; Berendse and Groenewegen 1991; Canteras et al. 1990; Parent and Hazrati 1995a, b).

There exists a species difference in the cortical areas projecting to the STN. In rodents, primary motor cortex, prefrontal cortex areas, anterior and medial cingulate cortex, agranular insular and the primary somatosensory region are involved, whereas in monkeys they are primary motor cortex (area 4) and premotor cortex (areas 6, 8 and 9). Moreover, some studies indicate that the whole cortex projects bilaterally (Rouzaire-Dubois and Scarnati 1985), others only ipsilaterally (Fujimoto and Kita 1993).

One should note that the bilateral projection in rat (Rouzaire-Dubois and Scarnati 1985) is also found in the lower primate Tupaia (Broere 1971), while ipsilateral projections are also described for Kalong and Capybara (Broere 1971). Seemingly there is strong species variability.

The cortical-subthalamic connections in man have also been studied by antidromic stimulation as caused in deep brain stimulation. One of the possibilities is that the effect of deep brain stimulation "may result from the spread of current to large-fibre systems near the STN" (Ashby et al. 1999) or the "activation of corticofugal (and

Table 1 Recalculation of the fibre diameters from Häggqvist sections through the cerebral peduncle at the subthalamic level

Medial region (Arnold's tract)	Intermediate area	Lateral region (Türck's tract)	Calibre
Frontopontine region	Cortico-spinal	Parieto-temperopontine	Häggqvist × 1.3
80%–95%	70%	85%/95%	0–1.3 μm
8%–15%	15.3%	8%–12%	2.6 μm
1%–5%	10.4%	1%–3%	3.9 μm
	3.2%		5.2–7.8 μm
	0.7%		7.8–10.4 μm
	0.3%		10.4–13 μm
	0.1%		>13 μm

not necessarily corticospinal) fibres could potentially generate potentials at the cortex through either of these routes" (Ashby et al. 2001). The velocities found should relate to fibre calibres of 4–5 μm or more (Ashby et al. 1999).

Using the Häggqvist method, Lankamp (1967) has studied the calibres of the corticofugal system. Counts were performed both at the level of the subthalamus/substantia nigra in the cerebral peduncle and the pyramid level just below the pons. Häggqvist fibre calibres can be re-counted according to those obtained with Feirabend and colleagues' electron microscopical techniques (Feirabend et al. 2002: factor is 1.3).

Recalculation shows that in the medial region (Arnold's tract; frontopontine corticofugal fibres) and in the lateral part (Türck's tract; parieto-temperopontine corticofugal fibres) all fibres present are below 4 μm (see Table 1). The intermediate area that contains the corticospinal fibres possesses only 4% of fibres larger than 5 μm. The total number of fibres larger than 5 μm stays constant compared to the lower pyramid counts. Both thick and thin fibres disappear in the corticospinal area in cases of amyotrophic lateral sclerosis (Lankamp 1967).

Therefore, corticosubthalamic fibres do not need to have a calibre of 4–5 μm or more. If in humans a contribution of the frontal or parietal cortex is involved, the fibres of the corticosubthalamic connection will certainly belong to the calibres below 4 μm. In fact, the human corticofugal system contains 85%–95% fibres smaller than 3 μm.

The first ultrastructural identification of corticosubthalamic axon terminals was presented by Romansky et al. (1979). They removed the primary sensorimotor areas in cats, and following 4–5 days survival looked in the STN. Only a moderate number of degenerating synaptic boutons (d.s.b.) were encountered in the ipsilateral STN. The d.s.b. contained round, densely packed synaptic vesicles, and were terminated by means of asymmetrical membrane specializations on spines and small dendrites, quite infrequently on proximal dendrites and never on neuronal perikarya. Rather rarely the d.s.b. contacted vesicle-containing dendrites, always being in presynaptic position. The d.s.b. had a patchy distribution within the subthalamic neuropil and a size corresponding to the smallest terminals (less than 1 μm) in the normal STN: the "small round" (SR) bouton (Romansky 1982; Romansky and Usunoff 1987; see also Sect. 2.2, this volume). In addition, groups of degenerating unmyelinated axons and relatively thin myelinated fibres were observed crossing the neuropil in small bundles. These data indicate that both categories of neurons in the STN, projection cells and local circuit neurons, are cortically dependent in the cat.

In addition, Bevan et al. (1995) found that glutamate-enriched corticofugal fibres in the rat, by means of asymmetrical synaptic thickenings, contacted the dendrites and spines of the subthalamic neurons.

5.2.2 Mirrored Somatotopy in the Subthalamic Nucleus

A dual somatotopy in *Macaca fuscata* has been favoured by Nambu et al. (1997; see Fig. 16). Projections of the primary motor cortex fulfilled the somatotopy that is known for the moto-corticosubthalamic connections (see above). Injections into the supplementary motor cortex showed anterogradely labelled endings into the STN but now not in the dorsolateral, but instead in the dorsomedial part of the STN. In fact the somatotopy of the primary motor cortex was mirrored in the somatotopy of the supplementary motor cortex (Fig. 16). Centrally in the dorsal part of the nucleus the legs are localized, while on both sides the arm and on both sides of the arm the oral-facial representation was found. "The existence of the highly ordered somatotopical representations in the STN would help to elucidate

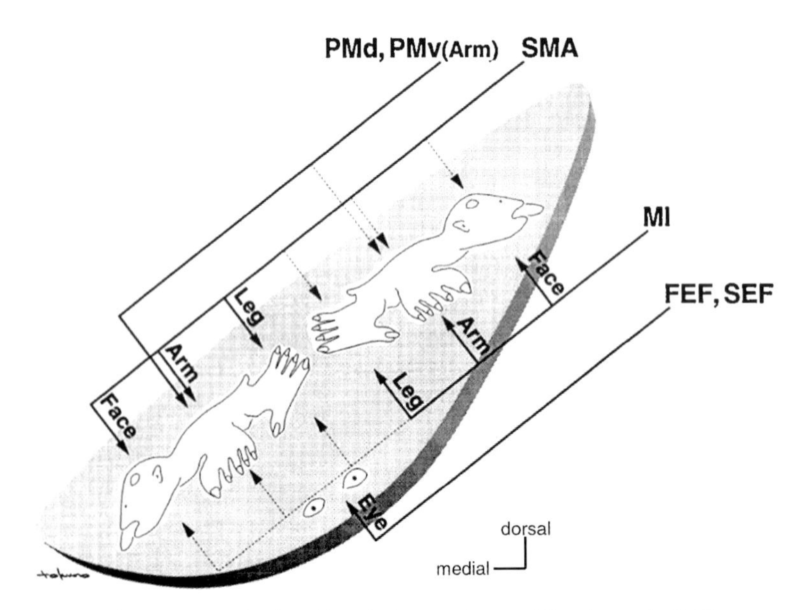

Fig. 16 Mirrored somatotopy in the STN of *Macaca fuscata* (Nambu et al. 1996)

how restricted lesions within the nucleus result in impaired movements of a single body part. No correlation in hemiballism has yet been revealed between the site of STN lesion and the somatotopical specificity of dyskinesia" (Nambu et al. 1997). Comparing the stippled figures of Nambu et al. (1997) concerning both types of cortical injections of the STN, a serious overlap is present for all three regions.

The somatotopic organization for the cortical afferents of the human STN has been studied in Parkinson's patients (Rodriguez-Oroz et al. 2001). Most neurons responding to leg movements were located laterally in the upper dorsal third of the STN, arm-related neurons were predominantly in the upper dorsal third and medial segment of the nucleus, while orofacial-related neurons were evenly distributed throughout the dorsal two-thirds, but mainly in the central portion (more ventral part) of the sensorimotor region. "Most likely these neurons represent those receiving afferents principally from the primary motor cortex, which is known to project to the dorsolateral STN in the monkey (Nambu et al. 1997)" (Rodriguez-Oroz et al. 2001).

In conclusion, the corticosubthalamic connections have not been proved in humans by neuroanatomical techniques. The motor corticosubthalamic connection is a moderate one in experimental animals as demonstrated in cats (see Romansky et al. 1979; Sect. 5.2.1, this volume) by electron microscopy and rat (Malinov et al. 1984). There exists strong species variability for the corticosubthalamic connection.

The motor corticosubthalamic somatotopy as described in man using electrophysiological methods is less sharp (Rodriguez-Oroz et al. 2001; compared to Künzle 1976; Künzle and Akert 1977; Macaca, Nambu et al. 1997), since arm, leg and orofacial neurons contain a serious overlap (Rodriguez-Oroz et al. 2001) of nearly 25%–30% in the latero-central-dorsal segment of the nucleus. The studied antidromic corticosubthalamic human connections seemingly do not fulfil the calibres needed for explanation of the velocities and latencies found.

5.2.3 Subthalamo-Cortical Connections

The horseradish peroxidase (HRP) technique can provide retrograde (perikarya) and anterograde (axons) labelling of neurons. Using HRP injections in the cortex of cats, retrogradely labelled neurons were found in the STN (Miyata 1986). Such a connection was already reported in the rat (Jackson and Crossman 1981). In rats the lateral half of the STN projects to the homolateral dorsal cerebral cortex. In cat the medial half of the homolateral STN projects to the frontal and temporal cortex. The presence of this *faint* connection is also supported by Fast Blue injections by Sloniewski et al. (1986). Subthalamotomy has been repeatedly carried out in Parkinson's patients (Guridi and Obeso 2001, and references herein). Although in several cases the location of the lesion has been controlled pathologically, no Nauta-degeneration post-mortem studies have been known to control degenerated connections. Therefore, in humans no neuroanatomical notion of such a connection has been published to date.

Electrophysiological stimulation of the human STN by single stimuli organizes a series of cortical-evoked potentials in the frontal and central regions. The latencies can be as short as 2–3 ms (Ashby et al. 2001; Baker et al. 2002; Hanajima et al. 2004). These results in Parkinson's patients could support the presence of a direct subthalamo-cortical connection in humans, although antidromic stimulation of a cortical-subthalamic connection is held responsible (Hanajima et al. 2004).

5.2.4 The Pallido-Subthalamic Connection

Terminology of the basal ganglia is inconsequent and frequently misused (Mettler 1968). Therefore the definitions used here are: striatum contains putamen and caudate nucleus, corpus striatum encompasses the putamen, caudate nucleus and globus pallidus, while lentiform nucleus comprises the globus pallidus and the putamen.

Kölliker (1896) found in his studies that the caudate nucleus and putamen are connected to the STN. Moreover, Von Monakow (1895) demonstrated a human case in which due to cortical lesions the postero-lenticular part of the internal capsule was missing. The putamen, caudate nucleus and globus pallidus stayed intact. From the striatal body, massive degeneration reached the intact STN. Thus, the older literature, as exemplified by two authors, considered the whole striatal body (putamen, caudate nucleus and globus pallidus) to contribute to the afferents of the STN. This idea that the putamen and caudate nucleus also contributed to afferents of the STN (the fibrae strioluysiannae) was supported by Papez (1942). That the pallidum externum was preferentially connected to the STN had already been discovered by Vogt and Vogt (1920) in man and supported for the pallido-fugal system by Morgan (1927).

The answer was given by animal experiments around 1950. The work of Marburg (1946) showed that all fibres from the caudate nucleus and putamen ended in the pallidum and none of these fibres traversed the nucleus. The statement is incorrect, because striatal fibres also end in the substantia nigra (Morgan 1927; Szabo 1962, 1967, 1970,1972; Nauta and Mehler 1966), but it closed the discussion on the contribution of the striatum to the STN. The fibre connections between the globus pallidus and STN were supported by the study of Martinez (1961). Whittier and Mettler (1949) showed that pallidal lesions brought degeneration in the subthalamic nucleus whereas lesions of the STN showed degeneration in the globus pallidus. Verhaart (1950) gave, also based on his comparative results, the correct answer: striatal connections (putamen and caudate nucleus) terminate in the globus pallidus. Two sets of fibres pass along the globus pallidus to end in the substantia nigra. Moreover, no striatal fibres reached the caudal mesencephalon.

Interesting is Kodama's overview (1926, 1927) on the basal ganglia in the *Swiss Archive for Neurology and Psychiatry* because it is far earlier than the experimental results, although it follows from publications of Vogt and Vogt (1920) and Morgan (1927). He describes the connections correctly: the striatum is connected to the globus pallidus reciprocally, Kodama explicitly states that the STN receives fibres from the external and internal part of the globus pallidus. The primitive fundament of these connections, however, is the globus pallidus internus connection towards the STN, although the contribution of both globus pallidus parts is equal. Moreover, the globus pallidus is also connected to the tuber cinereum, especially the lateral hypothalamus (see however, Nauta and Mehler 1966). A crossed connection between both or either part of the globus pallidus and the STN is denied. The connections between globus pallidus and STN therefore are considered strictly homolateral. Moreover, a reciprocal connection between substantia nigra and globus pallidus externus was discerned.

Von Monakow (1895) subdivided the ansa lenticularis into a dorsal (known as fasciculus lenticularis), middle and ventral part (known as ansa lenticularis *sensu strictiori*). The middle part is usually indicated as the pallido-subthalamic tract. Gebbink (1967, 1969) showed that these divisions in humans are different trajectories for the same fibres. Pallidal fibres accumulate in three systems: the ansa lenticularis s.s., the lamina medullaris intima and the lamina medullaris accessoria. Together they form the pallido-fugal system (Fig. 17). The part of the ansa lenticularis beneath (ventral of) the globus pallidus gives off bundles that penetrate the posterior part of the capsula interna as the ventral part of Edinger's comb system.

The lamina medullaris accessoria is "a loose mass of fibres within the pallidus internus, which does not reach the capsula interna dorsally except for its medial tip. Only here it supplies fibres to the comb system. Medially, the distinction between the ansa and lamina medullaris accessoria no longer is feasible" (Gebbink 1969; Fig. 17A). Between the posterior part of the capsula interna and the globus pallidus a fibre layer called the lamina medullaris intima is present. "Also from this lam-

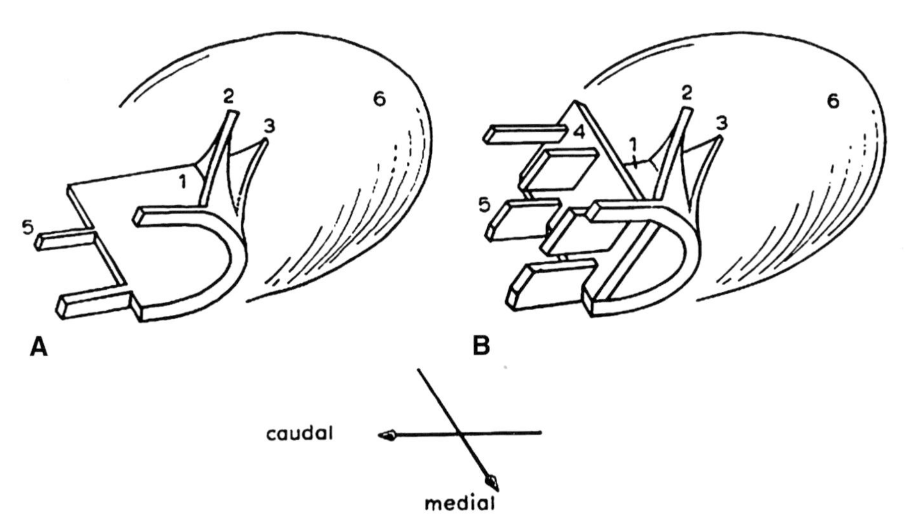

Fig. 17 Three-dimensional diagram of the pallidal systems within and around the pallidus internus viewed mediocaudally. **A** The ansa lenticularis *s.s.* (*1*), the lamina medullaris accessoria (*2*) and interna (*3*). **B** Also the lamina medullaris intima (*4*) is depicted. The ribbons of pallidal fibres (*5*) in the capsula interna are indicated. Globus pallidus (*6*). (Gebbink 1969)

ina ribbons of pallidal fibres detach to enter the crus posterius" (Gebbink 1969). All three systems in humans contribute to Edinger's comb system and only few pass medially around the posterior part of the capsula interna (Fig. 17B). These fibres pass into the field H of Forel. These pallidal fibres together with the striatal fibres constitute the comb system of Edinger. Within the posterior part of the capsula interna reorganization occurs: the striatal bundles detach from the pallidal ones. The striatal ribbons enter the substantia nigra (pars reticulata). The pallidal fibre bundles separate in a part that will constitute the fasciculus lenticularis dorsally and a part that is localized ventrally that course into the STN.

The fasciculus lenticularis is constituted of thin and thick calibre fibre bundles. The pallidal ventral thick ones pass into the STN. In the STN they spread into individual fibres. The smaller ones are restricted to the zona incerta.

"Von Monakow's division of the ansa lenticularis into a dorsal, middle and ventral division is misleading. His middle division, the fasciculus pallido-subthalamicus, actually consists of fibres derived from his ventral division, the ansa *s.s.* and the ventral part of the lamina intima, passing into the ventral part of the comb system" (Gebbink 1969). Thus, in humans a restricted area in ansa lenticularis that contains most pallido-subthalamic fibres is denied.

Based on fibre diameter, coarse fibres were studied in the basal ganglia. These pallido-fugal fibres were followed through the diencephalon by Verhaart (1950) who found these fibres to terminate in the STN. Using the Häggqvist method (for description see Appendix 2 and Marani and Schoen 2005) in pathological cases, the pallido-subthalamic connections were affirmed (Verhaart 1957).

Gebbink's thesis (1967) on the structure and connections of the basal ganglia in man contains a series of Häggqvist and Nauta-Gygax series that unquestionably confirm (presumably for the first time) the human pallido-subthalamic connection. The lesions of the globus pallidus internus and externus does not allow a good differentiation between both pallidal parts; nevertheless, "in globus pallidus lesions degeneration always is seen within the corpus Luysi in Häggqvist- and Nauta-stained sections alike" (Gebbink 1967). The course of the entering fibres from the capsula interna is from the rostroventral side (entrance) with a spread towards the caudomedial end. The second stream of entering fibres is those from the ansa and the comb system medially, which course caudo-lateralwards to enter the STN. Preterminal degeneration was found in the STN.

In 1969 *Psychiatria, Neurologia and Neurochirurgia* published an edition (vol. 72) in honour of the retirement of Verhaart as director of the Institute of Neurological Sciences at Leyden University. In this bundle of articles, Irena Grofová published an article on the topical arrangement of the pallido-subthalamic fibres in the cat. Using Nauta degeneration she found that

> ...the rostromedial part of the globus pallidus, which represents its most medial tip, projects on the most medial part of the STN, while rostrolateral and caudomedial parts of the globus pallidus, which are located at approximately the same distance to the midsagittal plane, project upon the central part, and the posterolateral part of the globus pallidus, which represents its most lateral tip, projects upon the most lateral part of the STN.

Moreover,

> the mediolateral representation of the globus pallidus in the nucleus subthalamicus is formed by almost horizontal layers. From their configuration it is evident that the lateral parts of the globus pallidus tend to project along the dorsal boundary of the STN.

This topology of the pallido-subthalamic connection has been proposed by Knook (1965) in the rat and by Nauta and Mehler (1966) in the monkey. The dorsoventral topology as proposed by Carpenter and Strominger (1967) in the monkey was refuted. The distinct organization in a mediolateral direction in the STN of the pallido-subthalamic fibres, however, shows a serious overlap in the antero-posterior direction. Therefore, the pallido-subthalamic connections were in principle solved by the Verhaart group (Verhaart, Knook, Gebbink, and Grofová) around the 1960s.

In 1966, at the founding of the journal *Brain Research*, the first article came from Nauta and Mehler on the projections of the lentiform nucleus in the monkey. The results concerning the subthalamic connections can be summarized as follows: it was Wilson (1914), using the Marchi method, who discovered the pallido-subthalamic connections. Heavy cell loss was noted in the STN by Vogt and Vogt (1920) from lesions of the external pallidal segment. Ranson and Ranson (1939, 1941; Ranson et al. 1941) using the Marchi method found renewed evidence for the pallido-subthalamic connections. The Nauta and Mehler (1966) paper supported the finding that the globus pallidus externus exclusively projects to the STN. However, the authors could not rule out that the internal pallidal segment will also contribute to STN. Moreover, the authors could not support the rostrocaudal topography of the pallido-subthalamic connections due to a lack of evidence.

The next series of articles directed itself to microcircuits between the globus pallidus and the STN. This refinement brought about another understanding of the loops involved in STN function and thus in hemiballism (see Smith et al. 1994, 1998).

A combined retrograde and anterograde study in the rat showed the dorsal pallidum (homologue to the globus pallidus externus) as the major source of afferent STN fibres (Canteras et al. 1990). A topographic organization of these projections was suggested, but could not be definitively ascertained. However, the combined anterograde and retrograde experiments indicated that "precisely organized feedback loops may exist" (Canteras et al. 1990).

Parent and Hazrati (1995a, b), in a total overview of the topographical organization of the pallido-subthalamic connection, concluded that the "lateral portion of the globus pallidus targets specifically the lateral two-thirds of the subthalamic nucleus, whereas those in the medial part of the globus pallidus and in the subcommissural ventral pallidum terminate respectively in the ventromedial and dorsomedial parts of the medial third of the STN" (Parent and Hazrati 1995a, b). Moreover, the rostrolateral region to which the globus pallidus externus (GPe) projects is also the region that projects back to the GPe. In the caudomedial area are located the STN neurons that project back to the GP internus. In experimental animals the topography of the pallido-subthalamic connections has been further elaborated by light and electron microscopical tracing (Smith et al. 1994; Shink et al. 1996). The reciprocal connections and topography is demonstrated in Fig. 18.

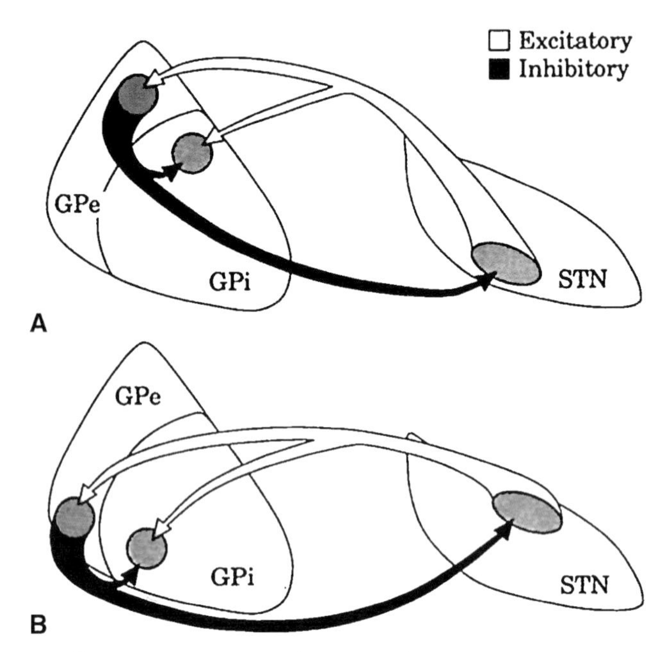

Fig. 18 Schematic diagram summarizing the relationships between the two pallidal segments and the STN. Small groups of interconnected neurons in the associative **A** and sensorimotor **B** territories of the GPe and STN innervate, via axon collaterals, a common functionally related region in the GPi (Shink et al. 1996)

The relations between the whole globus pallidus and STN are seemingly determined by small neuronal groups in squirrel monkeys (Smith et al. 1994). Groups in the GPe and STN innervate the same groups in the GPi. Medial and lateral parts of the GPi and GPe project to lateral and medial parts of the STN, respectively. Moreover, "individual neurons in the GPe project via collaterals to both GPi and STN. Similarly, the same population of neurons in the STN project to the same populations of neurons in the GPi and GPe" (Shink et al. 1996). These reciprocal connections are considered the anatomical substrate for the complex sequences of excitation and inhibition in the STN (Smith et al. 1994, 1998).

The ventral pallidum that encompasses "a region directly ventral to the anterior commissure, i.e. the subcommissural part of the ventral pallidum, and finger-like rostral extensions of this region into the deep layers of the olfactory tubercle" connects also to the STN (Groenewegen and Berendse 1990). The rat STN receives input from the ventral pallidum in a mediodorsal rim, just above the endings of the medial globus pallidus (Groenewegen and Berendse 1990). The projections are also related to the lateral hypothalamus, described as a homologue of the STN. Presumably the olfactory part of the ventral pallidum projects to this lateral hypothalamic area (Groenewegen and Berendse 1990). These connections are further elaborated in the rat (Berendse and Groenewegen 1991) in which the lateral STN receives input from the globus pallidus externus, and the medial ventrolateral part from the medial globus pallidus externus, while the lateral ventral pallidum projects on the medial-dorsal rim and the medial ventral pallidum on the lateral hypothalamic/STN region (Berendse and Groenewegen 1991).

Moreover, slight projections of the ventral striatum (putamen and caudate nucleus) were found "to enter the STN, indicating the existence of projections from the ventral striatum to this nucleus" in the rat (Groenewegen and Berendse 1990). Most of these projections end in the lateral hypothalamic area, adjacent to the STN. One should note that the STN in the rat is an open nucleus, spreading its dendrites far outside the STN (see Sect. 2.1, this volume). It is therefore probable that putamen and caudate endings do reach the STN dendrites.

In Grofová's diagrams (1969), one of the injections involves the cat lateral ventral pallidum, and indeed the medial dorsal rim of the STN contains Nauta-Gygax degeneration, while in Gebbink's series H5541, where the lesion touches on the ventral pallidum, the medial dorsal rim is also filled with degeneration. It therefore could well be that the results of Groenewegen and Berendse (1990; Berendse and Groenewegen 1991) also holds for cats and humans. Individual axons of the globus pallidus externus could be traced in primates (Sato et al. 2000a). These cells have demonstrated short intranuclear collaterals that branched near their cell bodies. Different patterns of targeting could be discerned, including those that project exclusively to the STN, others that project both to the STN and substantia nigra pars reticulata and neurons that projected to both globus pallidus internus and STN. These neurons ended on the STN cells with large varicosities at the soma and proximal dendrites. The single globus pallidus externus neurons can exert various effects on different targets among them the STN.

In conclusion, the pallido-subthalamic connections are well established in man. A pallido-subthalamic bundle is not restricted to within the ansa lenticularis as these fibres are found over the whole ansa. However, in humans there are only weak arguments for the point-to-point somatotopy as found in animal experimental research. Any contribution from the striatum has to be precluded in humans since the STN clearly is a closed nucleus.

The spread of the pallido-fugal fibres that *was thought* to be present in the cerebral peduncle changes into a clear pallido-peduncular bundle (bundle of Poppi 1927) at the start of the substantia nigra, according to older literature. The interaction of this bundle and the pars reticulata of the substantia nigra is peculiar. The relocalization of the so-called pallido-peduncular bundle out of the pyramidal part of the cerebral peduncle is characterized by an area of the pars reticulata that intermingles between the cerebral peduncle and the pallido-peduncular bundle (section 10, Figs. 19 and 20). The development of the pars compacta at the medial side of the pars reticulata does not change the situation that a part of the reticulata is in between cerebral peduncle and the pallido-peduncular bundle (section 20, Fig. 19). Although the pars reticulata slims its position in between the two structures, the bundle stays present (section 30–40, Fig. 19). The last reticulata cells are found in a triangular area between the cerebral peduncle and the ventro-lateral wall of the pallido-peduncular bundle. The bundle size also decreases (section 71, Fig. 19). At the level of section 90 the pars reticulata ends.

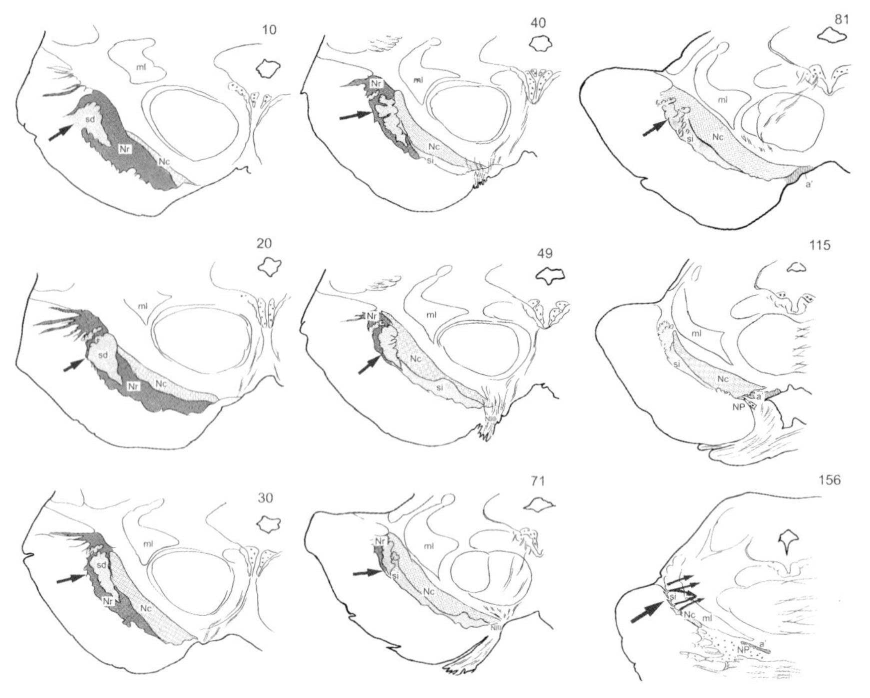

Fig. 19 Sections through the substantia nigra. *Arrow* indicates the position of the pallido-peduncular bundle. In section 154 the *arrow* shows the area where the pallido-peduncular bundle traverses the medial lemniscus (sections from series H3655, human, see Appendix 1)

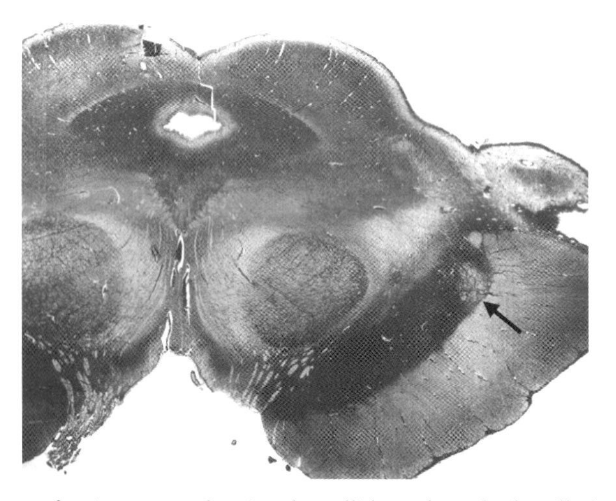

Fig. 20 Section 52 of series H3655, showing the pallido-peduncular bundle (*arrow*)

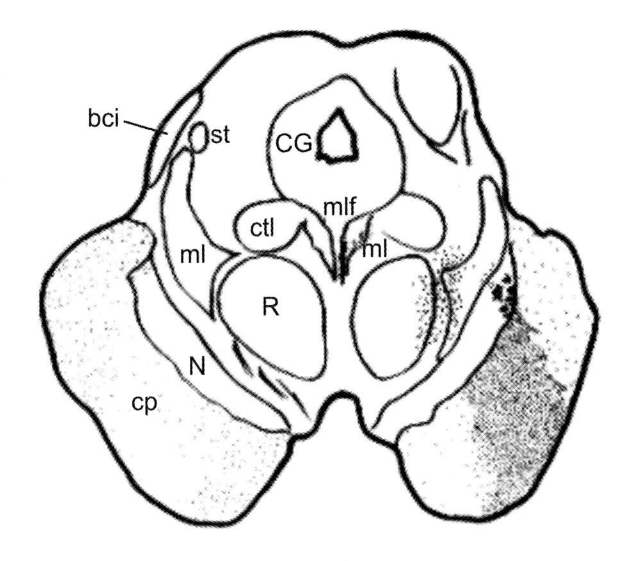

Fig. 21 Series H6348 (see Appendix 1): cortical lesions cause the degeneration of the pyramids. Degeneration related to the description was enhanced. For abbreviations see the abbreviations list

The lemniscus medialis will be crossed by the pallido-peduncular bundle (section 156, Fig. 19). This crossing can be followed in series H3655, due to the extremely good staining.

The pallido-peduncular bundle is filled with corticofugal degeneration from the pyramidal tract part of the cerebral peduncle (Fig. 21). The results from series H6348 (Fig. 21) show that terminal degeneration hardly penetrates above the area of the red nucleus-medial lemniscus.

Although Nauta-Gygax preterminal degeneration is difficult to interpret in these cases, seemingly only a minor innervation (or none) reaches into the pedunculopontine nucleus by fibres from the pyramidal or central part of the cerebral peduncle.

In series H5671 (Fig. 22, see Appendix 1) the whole cerebral peduncle is degenerated and gives off degenerated fibres filling the area between the medial lemniscus and the medial geniculate body. Clearly it can be noticed that the degenerated fibres pass as thick bundles through the medial lemniscus. The system diminishes since terminal degeneration is found in the nucleus paralemniscalis and the nucleus peripeduncularis. The area of the pallido-peduncular bundles in the dorsolateral part of the substantia nigra is massively degenerated. Diffuse preterminal degeneration is found, coming from these bundles in the cuneiform nucleus. In the nucleus pedunculopontinus tegmenti clear preterminal degeneration is found, as well as in the pars dissipata (see also Schoen 1969).

Moreover, immediately dorsal of the substantia nigra, a small elliptical field of densely packed, degenerated fibres of delicate medium size (indicated by Gebbink 1967, as the dorsal capsule of the substantia nigra) was noted. They extend rostrally into the STN that, however, proves to contain a split-shaped lesion in this series.

In conclusion, cortical bundles presumably can reach the STN via the pallido-peduncular bundle (better called the corticomesencephalic bundle) or dorsal capsule

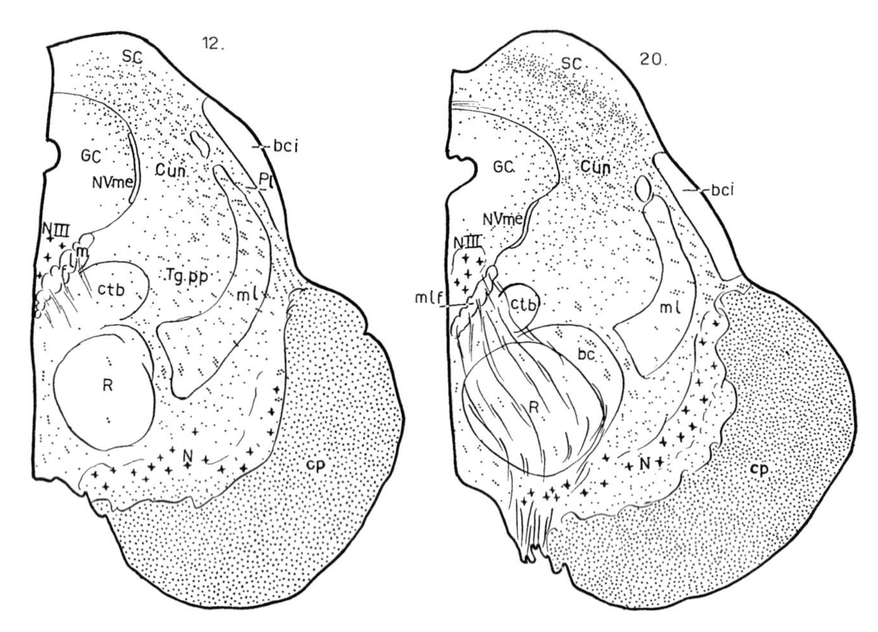

Fig. 22 Sections 12 and 20 of series H5671. Degeneration of the whole cerebral peduncle reaches the pudunculopontine nucleus. Compare with the picture from series H6348. For abbreviations see the abbreviations list

of the substantia nigra and halfway give off preterminal degeneration to the pedunculopontine nucleus.

In the cat, frontal cortical projections to the midbrain tegmentum have been described by Grofová (1965). The course of the degenerated fibres of cortical areas 4, 6 and 8 is through the substantia nigra, and the fibres terminate, among other nuclei, in the nucleus pedunculopontinus. The majority of these corticomesencephalic fibres are uncrossed, except for those passing through the periaqueductal grey and in the raphe.

Hirasawa et al. (1938) coagulated cortical area 22 in monkeys (*Pithecus fascicularis*) and studied them using the Marchi method. In general the same pathway to the mesencephalon for degenerated fibres was described, even the passage through the lemniscus medialis. However, endings in the STN were not found. As compared to series H6348 and H5671, the lateral part of the cerebral peduncle (parieto-tempero pontine part) demonstrates a clear preterminal degeneration in the pedunculopontine and presumably in the subthalamic nuclei. This is even more supported by case H5747 (see Appendix 1), where the temporal-cortical degeneration is found in the pedunculopontine nucleus. Series H5889 (see Appendix 1) seemingly indicates that the frontopontine part of the cerebral peduncle reaches the tegmentum of the mesencephalon. Therefore the general description as given by Voogd et al. (in Nieuwenhuys et al. 1988) that fronto- and parieto-temperopontine bundles will produce fibres that reach the mesencephalic tegmentum also holds for man. Both parts reach the pedunculopontine nucleus, and presumably these fibres could also reach the STN.

5.2.5 The Pedunculopontine-Subthalamic Interconnections

In the 1930s, Verhaart (1938a) denied the extrapyramidal connections over the red nucleus in his comparison concerning the corpus striatum and the red nucleus as the subcortical centra of the cerebral motor system. Moreover, his group showed that the rubro-spinal tract in humans can be neglected compared to other experimental mammals (Verhaart 1938b; Schoen 1964; see also Voogd et al. in Nieuwenhuys et al. 1988).

Nevertheless it was well known that mesencephalic stimulation influenced motor behaviour (Fuster and Uyeda 1962; Shik et al. 1966; Sterman and Fairchild 1966). Research showed the nucleus pedunculopontinus to be involved in locomotion, among other mesencephalic nuclei, especially in locomotion induction (see Garcia-Rill 1991; Inglis and Winn 1995, and references herein).

The localization and size of the pedunculopontine nucleus (PPN), however, is debated. Those that advocate an extended localization of the PPN (Garcia-Rill 1991) expand the nucleus from the posterior end of the substantia nigra to the laterodorsal tegmental nucleus.

Inglis and Winn (1995) extensively reviewed the literature on the localization of the PPN. There is also a lesser known painstaking review by Usunoff et al. (2003). It is now generally accepted that the PPN contains two groups of neurons: cholinergic and non-cholinergic neurons. The cholinergic ones concentrate in the centre of the reticular formation at pontine-mesencephalic tegmental levels, with two arms of cholinergic neurons embracing the non-cholinergic part of the nucleus. Non-cholinergic neurons always intermingle with cholinergic ones (Spann and Grofová 1992). There is growing evidence that the excitatory neurotransmitter glutamate is also present in the PPN: both cholinergic and non-cholinergic populations should contain glutamate (see Clements and Grant 1990; Clements et al. 1991; Lavoie and Parent 1994a, b; Ichinohe et al. 2000; Grofová and Zhou 1998), Moreover, an unexpected result is that GABA is also present in nearly half of the cholinergic neurons (Jia et al. 2003). These cholinergic neurons also contain NO (Vincent et al. 1983; see also Usunoff et al. 2003 and references herein). Other neurotransmitters present are catecholaminergic (Jones and Beaudet 1987) and peptidergic neurochemical markers (Vincent 2000).

The nucleus is located close to the ascending cerebellar superior peduncle in an area "bordered anteriorly by the substantia nigra, posteriorly by the parabrachial nucleus, dorsally by the cuneiform nucleus and deep mesencephalic nuclei, and ventrally by the pontine reticular nucleus" (Inglis and Winn 1995). However, studies that restricted the PPN to pure cholinergic neurons (Rye et al. 1987) and that considered a midbrain extrapyramidal area (MEA) of non-cholinergic neurons (Rye et al. 1988; Lee et al. 1988) to exist, attracted serious attention.

The cholinergic groups Ch5 and Ch6 as defined by Mesulam et al. (1983) can be related to the neuroanatomically defined nuclei in this area: Ch5 coincides with the

PPN and subpeduncular tegmental nucleus, while Ch6 is identical to the laterodorsal tegmental nucleus (Inglis and Winn 1995). Others take the subpeduncular tegmental nucleus into the PPN (Woolf and Butcher 1986), subdividing the PPN into a dorsal and a ventral part. Moreover, the NO-positive cholinergic neurons can be localized with NADPH diaphorase; in the rat, Usunoff et al. (2003) showed that its pars dissipata intrudes into the substantia nigra. Therefore the borders of the substantia nigra towards the PPN, but also towards the STN, are difficult to estimate. These arguments on the subdivision and borders of the PPN have consequences for the PPN-STN connections, since in human material only the Ch5 and Ch6 groups can be localized unequivocally (Mesulam et al. 1989). The older literature sometimes incorporated the PPN into the sub-cuneiform nucleus. So, PPN-STN connections are difficult to interpret in human material, without well-defined terminology.

The oldest subdivision discerns a pars compacta and a pars dissipata of the human PPN (Jacobsohn 1909; Olszewski and Baxter 1954). The pars dissipata, supposed to contain glutamatergic neurons, contains spread cells in the medial lemniscus and the cerebellar superior peduncle (Mesulam et al. 1983; Geula et al. 1993; Lavoie and Parent 1994a, b). Both parts contain cholinergic neurons. In humans the pars compacta contains 90% cholinergic cells and the pars dissipata contains 25%–75% of them (Mesulam et al. 1989).

> Although the cholinergic neurons in the lateral pontomesencephalic tegmentum are concentrated in the PPN pars compacta, they are intermingled with a variable portion of non-cholinergic neurons and penetrate into the territory of anatomically well-defined nuclei. Therefore, the definition of the PPN as consisting only of large, cholinergic neurons (Rye et al. 1987; Lee et al. 1988) is hardly tenable. Similarly, the concept of a non-cholinergic midbrain extrapyramidal area (MEA) (Rye et al. 1987; Lee et al. 1988) did not receive support from anatomical and physiological studies. (Spann and Grofová 1992).

It seems better to use Olszewski and Baxter's definition (1954), since all human-based descriptions use it, which makes comparison with the cytoarchitecture possible, especially in electrophysiological locations of electrodes, and it overcomes the confusion of "cytoimmunochemical versus Nissl localization" that appears in the literature. Moreover, modern publications (Pahapill and Lozano 2000) use Voogd's atlas of the brainstem in Nieuwenhuys et al. (1988), which in fact is a cytoarchitectural atlas, also based on Olszewski and Baxter, as is the atlas of Paxinos and Huang (1995).

In a subdivision of the pontine-mesencephalic reticular formation, Lakke (1997) brought the PPN into the nucleus reticularis cuneiformis. From E18 onwards, outgrowing fibres from the cuneiform nucleus are found in the rat's cervical intumescence, descending to lower spinal cord levels from E19 until P4. Moreover, the PPN neurons are born on E13 (Phelps et al. 1990). Therefore, the PPN is developing from E13 until P4, and its descending connections seemingly show their typical mature appearance after P5.

Based on the immunohistochemical localization of NOS, a few positive cells were found at E15 and the distribution pattern was completed by E19 (Terada et al. 2001). The cells of the PPN double-fold their average diameter at 2 weeks after birth, while at 5 weeks after birth this increase is reduced by half (Skinner et al.

1989). This change in diameter was reinvestigated by Kobayashi et al. (2004) and they found that this change exclusively occurred in the cholinergic compartment of the PPN. The development of the ascending projections of the cholinergic part of the PPN is slow, and the projections do not reach their mature targets before P28 (see Carden et al. 2000; Kaiya et al. 2003).

Experimental evidence shows the PPN-STN connection to be bilateral in cat and rat (cat: Graybiel 1977; McBride and Larson 1980; Nomura et al. 1980; Moon Edley and Graybiel 1983; Romansky and Usunoff 1983; Woolf et al. 1990; rat: Hammond et al. 1983c; Jackson and Crossman 1983; Takakusaki et al. 1996; Ichinohe et al. 2000; Orieux et al. 2000; ; squirrel monkey: Lavoie and Parent 1994a, b). Restricted autoradiographic injections in the PPN proper showed labelling over the *whole* STN, while HRP injections in the STN demonstrated labelled neurons around the cerebellar superior peduncle (Nomura et al. 1980). The homolateral projection is far more extensive than the contralateral connection.

In the rat an analogous result was found (Saper and Loewy 1980; Hammond et al. 1983c; Jackson and Crossman 1983; Sugimoto and Hattori 1984; Rye et al. 1987; Lee et al. 1988). The connections were found to be a reciprocal connection, which could be ascertained with neurophysiological methods (Hammond et al. 1983a, b). Due to the velocity (1.7 μ/s), it was concluded that the pedunculopontine fibres are unmyelinated. Moreover, the PPN-STN connection is a collateral connection from the PPN entopeduncular/globus pallidus connection (Hammond et al. 1983a, b). The rat PPN sends cholinergic, glutamatergic and GABAergic projections into the STN (Bevan and Bolam 1995). The bilateral connection of the PPN towards the STN has also been confirmed in primates (Carpenter et al. 1981; Lavoie and Parent 1994a, b). Their origin from cholinergic or from non-cholinergic neurons is unclear (Woolf and Butcher 1986, cholinergic; Lee et al. 1988, non-cholinergic), including from what parts of the PPN this connection originates (Pahapill and Lozano 2000).

The STN-PPN connections have been demonstrated by Carpenter and Strominger (1967), Nauta and Cole (1978), Carpenter et al. (1981), Jackson and Crossman (1981), Moon Edley and Graybiel (1983), Kitai and Kita (1987), and Steininger et al. (1992). Most are rat studies and the connection is considered glutamatergic (Pahapill and Lozano 2000) and thus excitatory. Nevertheless, the ipsilateral connection from the STN towards the PPN is termed "contentious" (Inglis and Winn 1995). "The small size of the STN has made resolution of its connections difficult" (Inglis and Winn 1995). Moreover, "the different subpopulations of rat PPN neurons that serve as targets for the STN input have not been established" (Pahapill and Lozano 2000) and this connection has not been confirmed in primates (Pahapill and Lozano 2000).

Although glutamatergic connections are considered excitatory, the neurophysiological studies on the rat STN-PPN connection are contradictory; Hammond et al. (1983a, b) have been found them inhibitory and Granata and Kitai (1989) excitatory.

In humans the PPN is subdivided into a pars compacta and a pars dissipata (Jacobsohn 1909; Olszewski and Baxter 1954). The human Ch5 group has been identified (Mesulam et al. 1989). However, none of the connections between STN and PPN has been established with neuroanatomical or neurocytochemical methods in humans.

5.2.6 Pedunculopontine Connections in Man

In Schoen's inheritance of human pathological cases studied using the Nauta degeneration technique and Häggqvist staining (for the descriptions of these techniques see Appendix 2; Usunoff et al. 1997; Marani and Schoen 2005), some of the series add to our knowledge of the connections of the pedunculopontine nucleus, dealt with in this section.

Marani and Schoen (2005) republished the series H5671, first published in Schoen (1969), showing the pyramidal degeneration in the human brain stem using Nauta-Gygax degeneration stain. The nucleus tegmenti pedunculopontinus receives minor cortical pyramidal degeneration from bundles that cross the medial lemniscus (see also Sect. 5.2.4, this volume).

In H5747 (see Appendix 1) at the level of the inferior colliculi (Fig. 23, section 46), similar to the rostral pons (Fig. 23, section 57), preterminal degeneration is found in the reticular formation lateral to the decussation of the brachium conjunctivum in the nucleus tegmenti pedunculopontinus and even between this decussation and the medial lemniscus, but the inferior colliculus does not contain any degeneration. In section 21 (Fig. 23) at the level of the superior colliculi, bundles of degenerated fibres detach themselves from the dorsolateral corner of Türck's tract to enter the tegmentum along the lateral aspect of the medial lemniscus. Termination occurs in the nucleus paralemniscalis between the medial lemniscus and the brachium of the inferior colliculus. Some fibres traverse the dorsal tip of the medial lemniscus to terminate in the nucleus cuneiformis immediately medial to it. The superior colliculus on the side of the cerebral peduncle contains abundant terminal and preterminal degeneration predominantly in its dorsal half. This degeneration proves to reach this colliculus through the brachium of the colliculus superior, as can be seen in section 10 (Fig. 23), slightly more rostrally. Besides this, a small number of degenerated fibres leaves Türck's area, and along the lateral aspect of the medial lemniscus reach the superior colliculus, giving off terminals to the nucleus paralemniscalis.

The ascending degeneration at the other side, found in the medial lemniscus, occupies the dorsal part of the latter. More rostrally in the midbrain, the medial lemniscus changes its horizontal position gradually into a vertical one, and the degeneration subsequently takes a more medial position (see also Marani and Schoen 2005).

It can be concluded that the human nucleus tegmenti pedunculopontinus receives pure homolaterally cortical temporal fibres.

In H6368 (see Appendix 1) the lesion of the oculomotor and trochlear nerve's nuclei, extending into the ventral half of the central grey matter, apparently causes terminal degeneration in its counterparts, as well as in the two-sixth nerve's nuclei. Although the mesencephalic lesion very evidently severs almost all of the mesencephalic tegmentum including the area of Forel's fasciculi tegmentales, retrograde degenerative changes are seen neither in Wallenberg's tract, nor in the dorsal portion of the main sensory trigeminal nucleus.

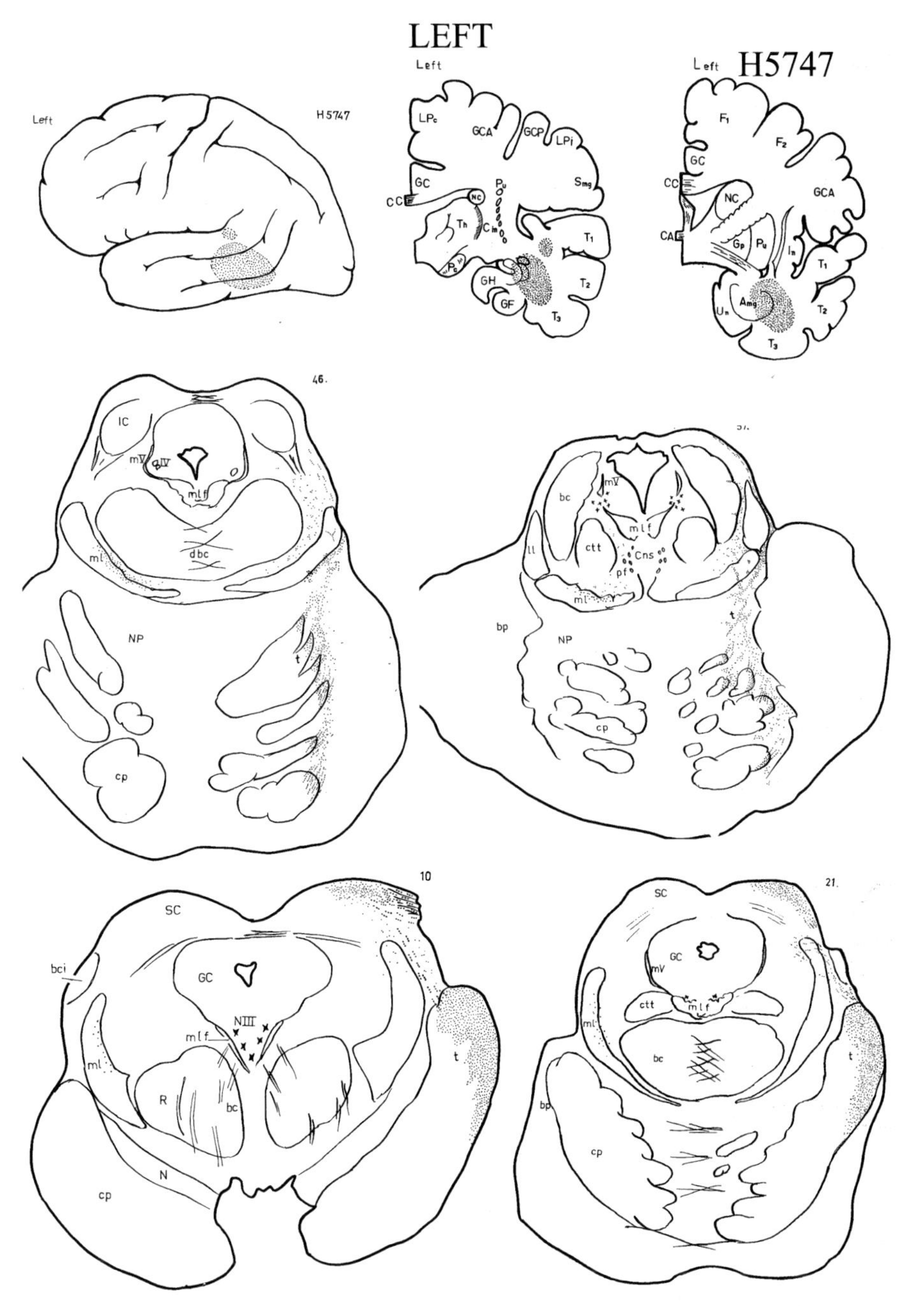

Fig. 23 Sections through series H5747. *Upper part* indicates the lesioned area in the cortex. *Lower part* demonstrates degeneration (*stippling*) from pons to mesencephalon. For abbreviations see the abbreviations list

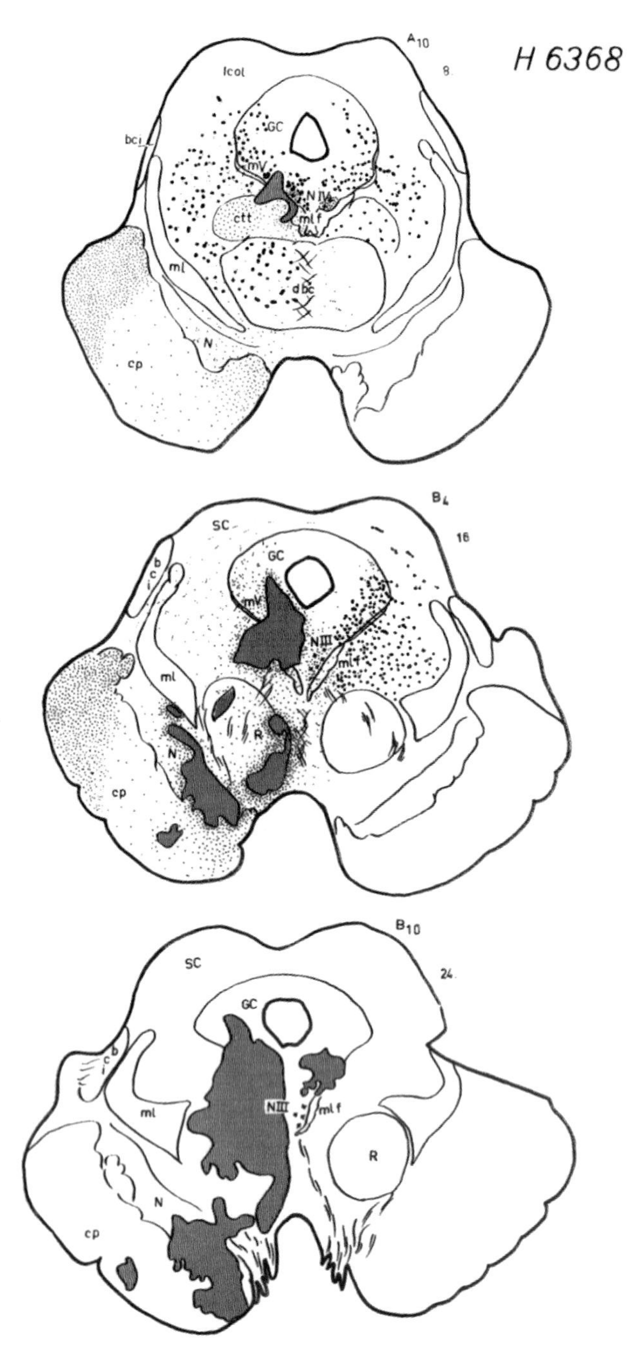

Fig. 24 Sections of series H6368. Degeneration as described in the text is enhanced. For abbreviations see the abbreviations list

A small lesion in the ventral portion of the central grey matter of the *right* half of the midbrain (Fig. 24), just touching the dorsolateral tip of the medial longitudinal fascicle, but sparing the oculomotor nerve nucleus, gives rise to considerable degeneration caudally at the same side. Some of these fibres leave the central grey, traversing the trigeminal mesencephalic root to terminate in the ipsilateral cuneiform and sub-cuneiform nuclei of the dorsolateral tegmentum. More caudally this degeneration can be found back in the area of the nucleus coeruleus, extending slightly into the sub-ependymal grey matter; caudal to the motor nucleus of the fifth nerve it is no longer present. The ventral central grey matter projects to the sub-cuneiform nuclei, among them the pedunculopontine nucleus, and spreads its terminal degeneration over the whole dorsolateral mesencephalic tegmentum. This type of degenerational localization has been extensively described by Hamilton and Skultety (1970) and Nauta (1958) in the cat, and is thus confirmed by this case.

In series H5889 the area around the brachium conjunctivum—medial, lateral and in the bundle—terminal degeneration was found. The degeneration stays sharply homolateral, exclusively in the bundle of the brachium conjunctivum descendens.

Seemingly contralateral projections from the ipsilateral tegmentum into the contralateral tegmentum are minimal, indicating that the connections of both pedunculopontine nuclei are absent or minimal.

The degeneration of both pyramidal tracts, one very old (thus unnoticeable with the Nauta technique) and the other more recent, follow the pathway described above, passing through the lemniscus and entering with a large spread into the tegmentum.

5.3 Raphe Connections to Subthalamic Nucleus

The serotonin connections were for the first time described by Steinbusch (1981), who used antibodies against serotonin. He discovered that formalin coupling of serotonin to a protein carrier produced antibodies to serotonin. Numerous positive dots were found into the STN, indicating a heavy labelling of the STN by serotoninergic terminals in the rat. A seemingly extensive projection from the raphe nuclei towards the STN was presented. Further elaboration by tract tracing and immunocytochemistry revealed the parts of the raphe nuclei that projected towards the STN. However, such a distribution is seemingly absent in parts of the opossum STN (Martin et al. 1985).

Vertes and colleagues (Vertes 1991; Vertes and Kocsis 1994; Vertes et al. 1999) subdivided the serotonin projections into a midline/paramidline area and a more lateral area, originating respectively from the median raphe and the dorsal raphe nucleus. There was no overlapping in their projection areas. The dorsal raphe nucleus projects to the STN (Bobillier et al. 1976, cat, with autoradiography; Mori et al. 1985, rat, cat and monkey; Lavoie and Parent 1990, squirrel monkey, with Mori et al. and Lavoie and Parent using immunocytochemistry). The serotoninergic effect is mediated by various types of 5-HT receptors present on the STN neurons (see Sect. 2.3.4.6, this volume). mRNA is present for several 5-HT receptors in the STN neurons in various concentrations (high concentration e.g. for 5-HT$_4$ and 5-HT$_{2C}$ receptors, Pompeiano et al. 1994). The serotoninergic input to the STN modulates the spontaneous firing of the rat subthalamic neurons by a decreasing potassium conductance by activating 5-HT$_4$ and 5-HT$_{2C}$ receptors (Xiang et al. 2005). Within the human STN intermediate concentrations of 5-HT$_7$ receptors were found (Martin-Cora and Pazos 2004).

The serotonin terminal distribution showed an even density over the STN in rat and cat, whereas in the monkey a ventro-medial preference was found, while thick serotonin-positive fibres passed through the monkey's STN (Mori et al. 1985). In the squirrel monkey a dense, evenly distributed network of serotonin terminals was present (Lavoie and Parent 1990). Only in the rat was a projection from the STN into the dorsal raphe nucleus ascertained by Kitai and Kita (1987).

5.4 The Thalamo-Subthalamic Connections

The cat centre-median parafascicular complex was studied by Sugimoto et al. (1983) by injections of a mixture of tritiated l-leucine and l-proline. The STN was heavily labelled in its rostral and ventro/ventromedial parts. The labelling diminished in the middle part of the STN to disappear in its caudal part. After HRP injections in the cat, STN retrogradely labelled cells were found in the centre-median parafascicular complex. Repetition of these experiments in the rat proved the same connection (Sugimoto et al. 1983). Confirmation of this ipsilateral connection was also obtained by autoradiographic electron microscopy (Sugimoto and Hattori 1983).

A complementary termination for the parafascicular nucleus and centre-median of the STN was found in the squirrel monkey (Sadikot et al. 1992). Hyperactivity of the parafascicular neurons projecting to the STN was proved using mRNA for the first subunit of cytochrome oxidase in rats containing a 6-hydroxydopamine unilateral lesion. Hyperactivity of the STN could in part be explained according to authors by an increase in excitatory input from the parafascicular nucleus (Orieux et al. 2000). A small contralateral projection from the parafascicular nucleus to the STN was described by Castle et al. (2005) in the rat. Moreover, STN neurons could also reach the contralateral parafascicular nucleus. The rat STN neurons projecting to the ipsilateral globus pallidus and to the ipsilateral substantia nigra pars reticulate received parafascicular terminals (Castle et al. 2005). Using anterograde and retrograde labelling the parafascicular subthalamic connections were also confirmed in monkeys (Tande et al. 2006).

5.5 The Subthalamic–Central Grey Connections

There are only limited publications that found a projection from the STN into the periaqueductal grey (Kitai and Kita 1987; Smith et al. 1990b). The periaqueductal grey is definitely involved in somatomotor behaviour (see Carrive 1993). Injections of *Phaseolus vulgaris*-leucoagglutinin in the monkey STN brought forward projections to the periaqueductal grey, indicating that the STN projects further downward into the mesencephalon and pons (Smith et al. 1990b). This was also confirmed by Parent (1996).

5.6 The Colliculus Superior Connections

Redgrave's group proposed several parallel loops that are involved in basal ganglia motor output. The best example of multiple loops are those that originate in the colliculus superior, pass the thalamus, the caudate and putamen and, via the substantia nigra pars reticulata, reach the layers of the superior colliculus again (McHaffie et al. 2005). An abstract recently appeared that indicates connections of the superior colliculus and the STN:

> We have therefore conducted a series of anatomical experiments in a range of species (rat, cat and monkey) to examine the tecto-subthalamic projection in more detail. Injections of the neuronal tracers biotinylated dextran amine or *Phaseolus vulgaris* leucoagglutinin into the SC produced anterograde labelling of the STN in each species. For example in rats injections into the lateral intermediate layers produced a dense sheet of terminals predominately in the dorsal STN, while more medial injections produced only sparse anterograde labelling. (Coizet et al. 2007)

5.7 The Nigro-Subthalamic Connections

Until halfway through the 1990s the projection from the substantia nigra pars compacta into the STN was not completely described (Hassani et al. 1997). Retrograde labelling of the substantia nigra pars compacta was obtained by injecting Fluor gold into the STN. The labelled neurons could also be determined with tyrosine hydroxylase. Confirmation of the connection was obtained by biocytin antegrade tracer injections in the substantia nigra pars compacta. Branching of the nigral axons was manifold, while a terminal arborization of $400^{*} \times 250^{*} \times 150\ \mu m^3$ was described in the rat (Hassani et al. 1997). These results are in agreement with earlier findings (Björklund and Lindvall 1984; Brown et al. 1979; Campbell et al. 1985; Meibach and Katzmann 1979).

The dopaminergic connections in primates remained controversial until 2000 (Francois et al. 2000). The nigro-subthalamic connections were confirmed in *Cercopithecus aethiops* using Fluor gold injections in the STN. The connections were re-established using biotin dextran injections in the A8 and A9 areas. Labelled axons were found throughout the whole extent of the STN. Using tyrosine hydroxylase immunoreactivity the catecholaminergic origin was firmly established. These fine ramifications of tyrosine hydroxylase-positive branches and varicosities were also noted in human STN. These tyrosine hydroxylase fibres showed a reduction of over 60% in MPTP-treated monkeys and in parkinsonian brains (Francois et al. 2000). Dopaminergic innervation of the human STN was supported by articles from

Cossette et al. (1999), Francois et al. (2000) and Hedreen (1999). An overview of the dopaminergic innervation in the basal ganglia is given by Smith and Kievel (2000).

6
Nigro-Subthalamic Connections in the Rat

6.1
Introduction

The STN projection neurons are glutamatergic, excitatory, and heavily innervated by widely branching axons of the substantia nigra (SN) (see Sects. 5.1 and 5.2.10, this volume). Leucine-labelled fibres of the STN follow in their projections the laminar organization of the substantia nigra's pars reticulata (Tokuno et al. 1990). However, the nigro-subthalamic connection remained controversial (see Sect. 5.2.10, this volume) due to its incomplete description in various experimental animals. Although functional dopamine receptors are expressed in the STN (see Sect. 2.3.4.1, this volume), the direct modulation of subthalamic neurons by dopamine of the substantia nigra is controversial owing to the low density of dopamine axons in the STN (see Cragg et al. 2004). Renewed tracing research was therefore carried out in the rat. To date, only an ipsilateral projection has been found for the connections between SN and STN. Using BDA, the SN-STN connection has been studied again, and a bilateral projection was established.

6.2
Materials and Methods

6.2.1
Injections

Twenty female Wistar Albino Glaxo rats weighing 200–240 g were used. The animals were anesthetized with Hypnorm (0.3 ml/kg i.p.; 0.2 mg/ml fentanyl; Ceva, Paris) and Valium (1.0 ml/kg i.p. 5 mg/ml diazepam; Hoffmann-La Roche, Basel). All rats further received a subcutaneous dose of 0.1 ml atropine sulphate (500 μg/ml) to diminish mucous secretion into the tracheo-bronchial tree. After mounting in a Narashige stereotaxic frame in the flat skull position, biotinylated dextran amine (BDA; 10%, mw 10,000; Molecular Probes Europe, Leiden, The Netherlands) dissolved in phosphate buffer (PB; 0.1 M, pH 7.2) was injected unilaterally into the SN using a vertical approach. Stereotaxic co-ordinates were obtained from Paxinos and Watson's atlas (1996). Injections were made through silicon-coated glass micropipettes (Yu and Gordon 1994), and the BDA solution was freshly prepared for each injection. Pressure injections were made using a Picospritzer, and iontophoretic injections with a Midgard CS3 iontophoretic power source (3–5 μA pulsed DC, 5 s on/off for 30 min). At the end of each injection the pipette was held in place for 15 min to insure that the inject BDA was absorbed into the tissue, and that there was not a significant spread of the tracer within the pipette track. Survival time was 8–13 days. The rats were deeply re-anaesthetized with Nembutal (1.5 ml/kg i.p. 60 mg/ml sodium pentobarbital; Sanofi Sante, Maasluis, The Netherlands), and perfused transcardially with 100 ml of 0.9% saline, followed by 500 ml of 4% formaldehyde (Merck, Darmstadt, Germany) in water. Immediately prior to perfusion sodium nitrite (0.5 ml; 1% in water) and heparin (0.5 ml; 5,000 IE/ml; Leo Pharmaceutical Products, Weesp, The Netherlands) were injected intracardially. The brains were removed, rinsed in water for 4 h, and soaked in 10% sucrose in water overnight at room temperature. Serial sections were cut at a thickness of 40 μm on a Jung freezing microtome, and collected in PB.

6.2.2 Tracer Histochemistry

A commercial avidin-biotin-HRP complex (ABC) kit was used to visualize the BDA (Vectastain ABC Kit, Vector Laboratories, Burlingame, United States). The sections were soaked in PB containing 0.1% bovine albumin (fraction V; Sigma Chemical Co., St Louis, United States) for 30 min, and rinsed in PB for 30 min. Then the sections were incubated in the avidin-coupled biotinylated-HRP solution for 60 min on a shaker, and rinsed again in PB for 30 min. The reaction product was developed with 0.06% 3,3'-diaminobenzidine (Sigma Chemical Co., St Louis, United States) and 0.02% H_2O_2 in Tris buffer (0.05 M, pH 7.6) for 15 min. The sections were then rinsed in distilled water, mounted on chrome alum-subbed slides and dried overnight. The sections were counterstained with cresyl violet, dehydrated through graded ethanols, cleared with xylene, and cover slipped with Entellan (Merck, Darmstadt, Germany).

6.3 Results

6.3.1 Tracing Results

6.3.1.1 Appearance of Labelling

Large injections (series C5778, 5785, Fig. 25) were used to describe the overall projections of the SN. In series C5778 the injection is almost throughout the rostromedial and lateral extent of the SN, and involves the lateral SNr and SNc and the SNl, with some involvement of MRN neurons covering the dorsal surface of the SN. This injection resulted in labelling throughout STN both ipsilateral anterograde and retrograde (il) and contralateral antegrade (cl). This series will serve as the prototype for the description of the labelling observed in and around STN. The overall results

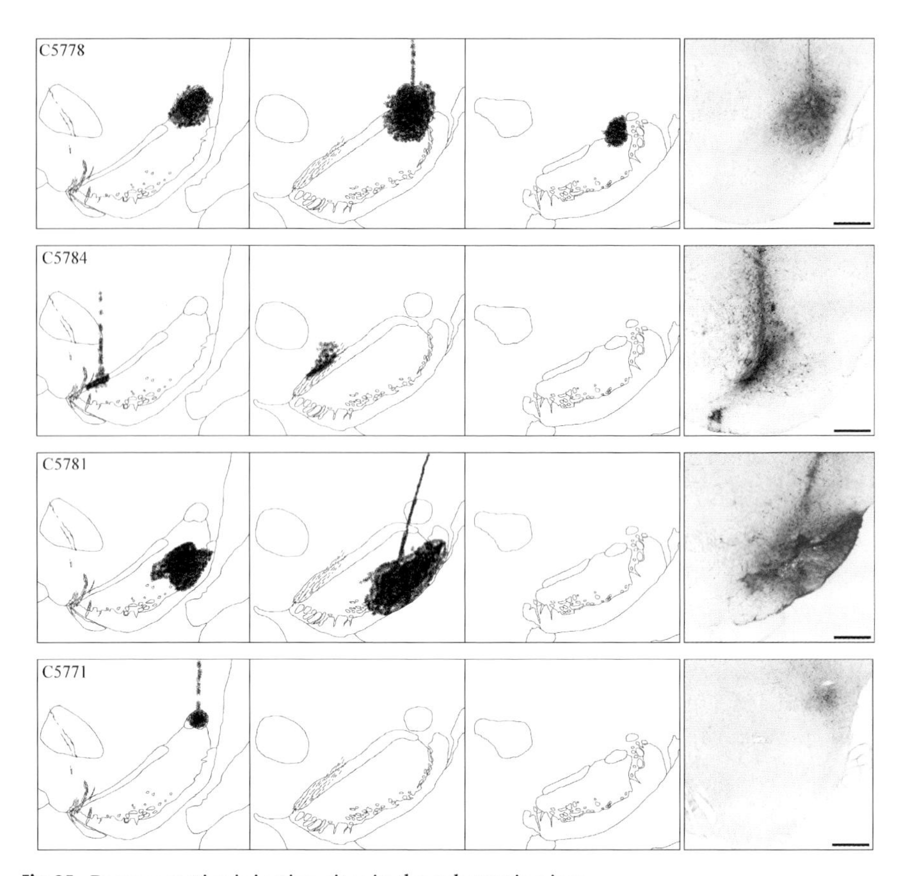

Fig. 25 Representative injection sites in the substantia nigra

from all series are shown in Table 2, while the characteristic STN connections are described separately.

Moreover, a series of injections just dorsal, rostrodorsal and rostral to the SN ($n = 8$) corroborated that the described SN connections originated in the SN.

6.3.1.2
Course and Termination of Nigrosubthalamic Connections

The largest number of nigrosubthalamic axons was observed in case 5778. The injection site of the tracer involved the mediolateral SNl, SNr and SNc (Fig. 25). The labelled axons radiate from the injection site. The axons are directed to the brain stem, and some nigrothalamic axons course dorsally towards the tegmentum, and the ascending axons to the forebrain initially take a medial course towards the prerubral area. Most of them run immediately dorsal to SN, and some axons traverse the SNc lateromedially. Few axons curve ventromedially and travel along the border between

Table 2 Connections of the substantia nigra

Afferents to substantia nigra							Efferents from substantia nigra					
SNr		SNc		SNl			SNr		SNc		SNl	
i	c	i	c	i	c		i	c	i	c	i	c
↳		↳		↳		Cortex	↲		↲		↲	
↳		↳		↳		Caudate-Putamen	↲		↲		↲	
↳		↳				Pallidum	↲		↲			
↳		↳				Accumbens			↲			
						Hippocampus			↲			
		↳		↳		Amygdala			↲		↲	
						Lateral dorsal thalamic nucleus	↲					
						Medial dorsal thalamic nucleus	↲	↲	↲			
						Ventral medial thalamic nucleus	↲	↲	↲			
						Central medial thalamic nucleus	↲		↲			
						Central lateral thalamic nucleus	↲			↲		
↳		↳		↳		Parafascicular thalamic nucleus	↲			↲		
						Paracentral thalamic nucleus	↲					
						Lateral posterior thalamic nucleus	↲					
↳		↳		↳		Lateral habenular nucleus			↲	↲		
						Dorsal lateral geniculate nucleus	↲		↲		↲	
						Zona incerta	↲					
↳		↳		↳		Subthalamic nucleus			↲			
↳	↳	↳	↳	↳	↳	Hypothalamus	↲		↲		↲	
						Superior colliculus	↲	↲			↲	
						Red nucleus	↲					
↳		↳				Entopeduncular nucleus						
						Inferior colliculus					↲	↲
						Periaqueductal gray			↲	↲		
↳		↳				Raphe dorsalis			↲			
						Cuneiform nucleus	↲					
						Mesencephalic reticular nucleus	↲		↲			
		↳	↳			Pedunculopontine tegmental nucleus	↲	↲	↲			
↳	↳	↳	↳	↳	↳	Laterodorsal tegmental nucleus	↲					
↳	↳	↳	↳	↳	↳	Parabrachial nuclei	↲		↲			
						Locus coeruleus	↲		↲		↲	
						Parvocellular pontine reticular nucleus	↲		↲			
			↳			Cerebellum		↲		↲		↲

i, ipsilateral; c, contralateral

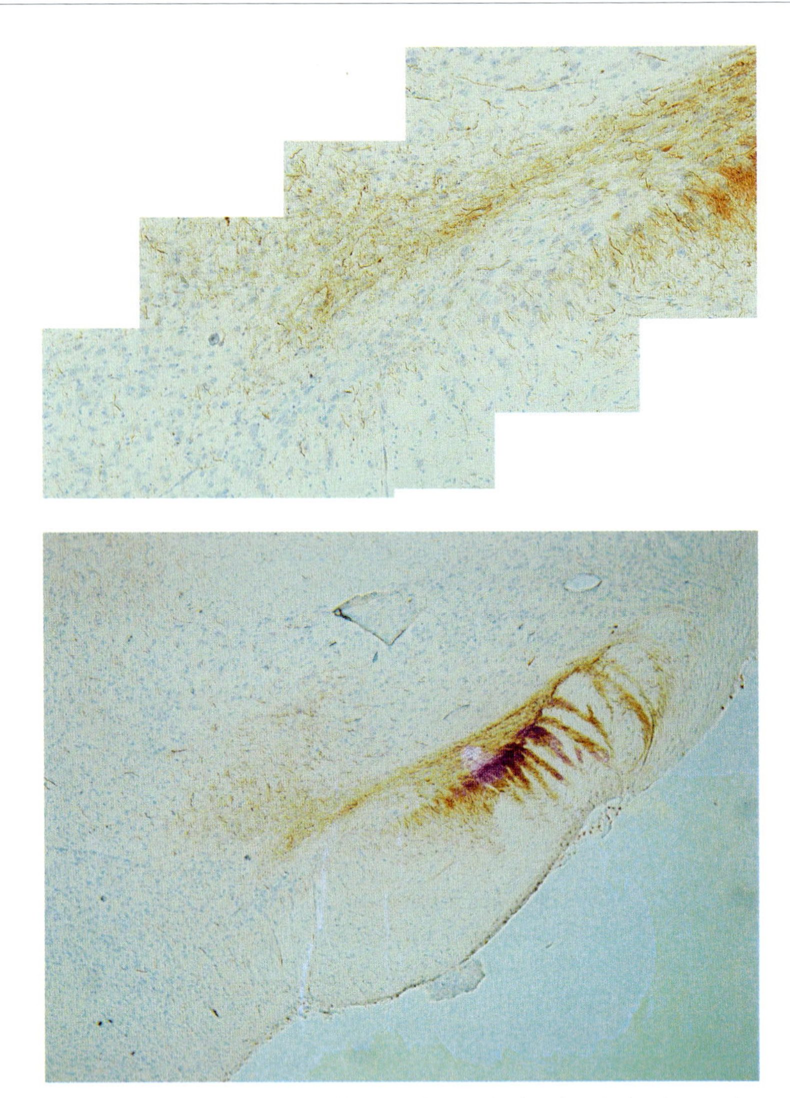

Fig. 26 Pathway of nigro-subthalamic fibres at the cerebral peduncle (series 5778)

SNr and the cerebral peduncle. Reaching the caudal pole of the STN (Fig. 26) the labelled axons enter the nucleus through its lateral wedge and from the medially running bundle, dorsal to the STN. Labelled axons also enter the STN through its ventral border, but their course is largely obscured by the numerous retrogradely labelled striatonigral axons, arranged in the bundles of the Edinger's *Kammsystem des Fusses* ("comb system of the foot"). Within the STN, especially in the lateral half of the nucleus, along with passing fibres oriented mediolaterally, there is a large amount of terminal labelling (Fig. 27). In the medial part of the STN there are mainly discrete bursts of terminal labelling. Interestingly, although the subthalamonigral projection

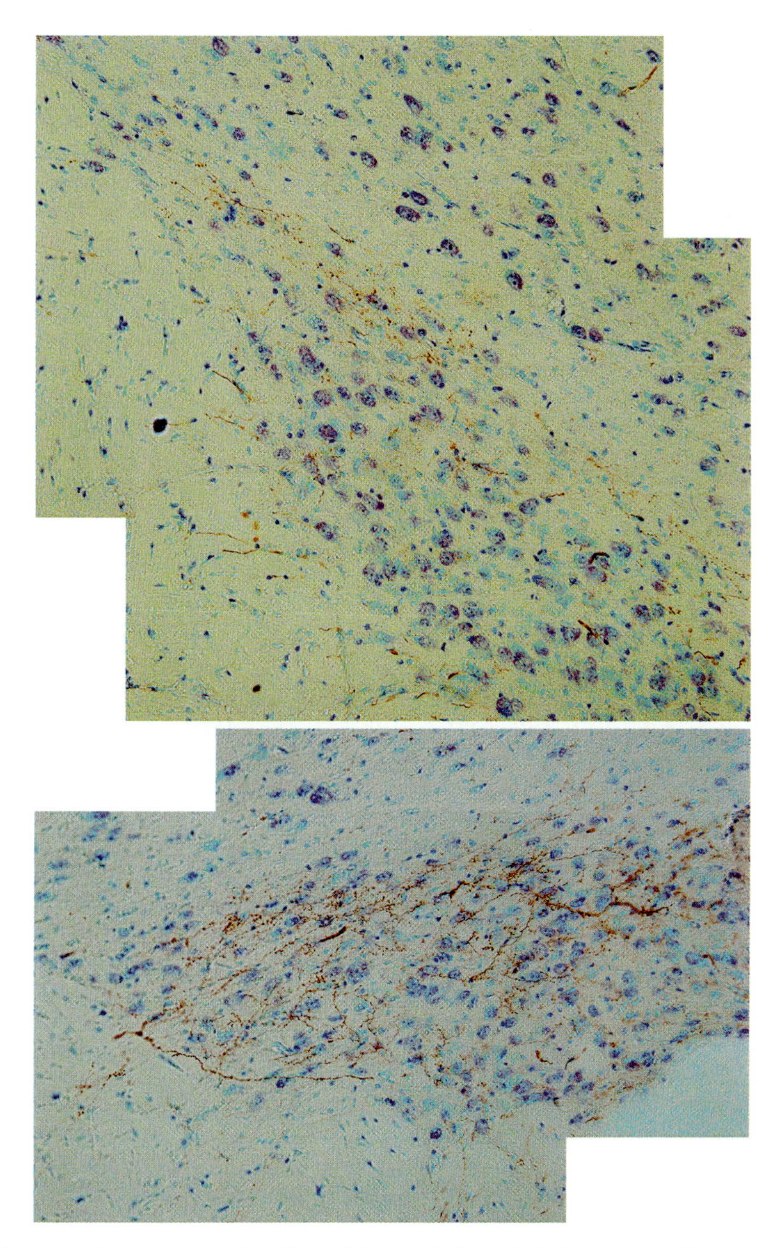

Fig. 27 Contralateral (*top*) and ipsilateral (*bottom*) nigrosubthalamic projections

is a substantial one, only few retrogradely labelled STN neurons are present, and most of them are not heavily loaded with the reaction product.

The SN axons cross the midline at several places. The most substantial component of crossed axons runs in the mesencephalic tegmentum ventral to the periaqueductal grey. Such bundles are present through the entire rostrocaudal extent of the mesencephalon, and some fibres in the rostral mesencephalon apparently enter the STN through its dorsal border. A second component crosses the midline in the commissure of the superior colliculus and in the posterior commissure. Although some fibres bend in the ventral direction contralaterally, none of these axons appears to enter the contralateral STN. Rostral to the SN, the efferent SN axons cross the midline in the adhesio interthalamica (crossed nigrothalamic axons), and the last component of crossing axons runs in the supraoptic decussation, immediately above the optic tract. Some of these axons take a dorsomedial course towards the contralateral STN. In the contralateral STN a considerably lower number of labelled axons are seen. However, they form very distinct mediolaterally extended patches that might be followed in serial sections. Most of these discrete fields of terminal labelling are in the central and lateral portions of the STN, but also medially some terminal "whorls" are seen.

6.3.2 Injections into the SNl

A small injection (series C5771) into the SNl is without further involvement of the SNc or SNr and is placed dorsolateral of the SNc. The ilSTN contains scant fibres entering the nucleus from the dorsolateral side, and few terminations are noticed around the cells of the STN. Labelled fibres and terminations are absent in the contralateral STN. An analogous but larger injection (C5838) in the SNl, reaching the SNc, shows, however, a strong termination pattern in the ilSTN and few labelled fibres in the cl STN. These fibres entered the nucleus from its laterodorsal side. An injection (C5830) just dorsal of the SNl into the peripeduncular nucleus and inferior colliculus shows absence of labelling in both the ilSTN and the clSTN.

6.3.3 Injections into the SNr

Nearly all injections involving the SNr also touched upon the SNc, which is due to the dorsal approach of the injections. Three series (C5606, C5785, C5778) contained large injections involving not only the SNr but the SNc too. In these three series, labelling was found in the ilSTN and clSTN. From the only two selective injections into the SNr (C5835, C5781) one was mainly placed in the cerebral peduncle (C5835). Passing cortical fibres and some nigral fibres are labelled, and these fibres find their trajectory over and through the rostral top of the STN. Terminal labelling and positive fibres are found in the ilSTN. The clSTN stays free from fibres and terminal labelling in C5835. However, in C5781, with a large injection in the SNr, both ilSTN and clSTN (Fig. 28) contained labelled terminals and fibres.

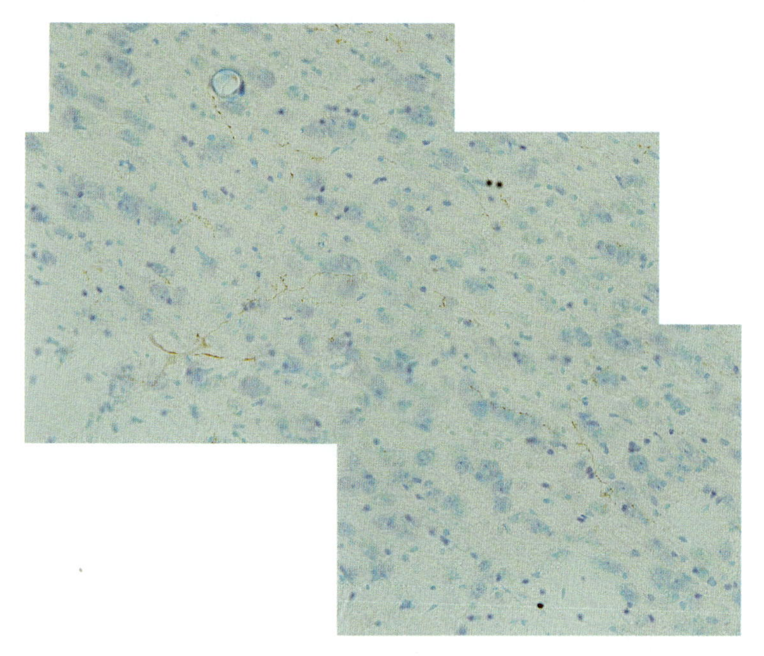

Fig. 28 Contralateral projection into the STN after SNr injection (5781)

6.3.4 Injections into the SNc

C5784 contains an injection, which is restricted to the medial part of the SNc. Heavy terminal labelling and labelled fibres are found in the ilSTN. A single retrogradely labelled ilSTN neuron was present. The clSTN contained sparse terminal labelling with a larger number of labelled fibres. The positivity was restricted to the caudal and lower middle part of the clSTN.

C5767 shows a small injection into the middle part of the SNc. Labelling (fibres and terminals) was noticed in the ilSTN. Due to the small injection size only faint terminal labelling and a few fibres are present in the middle of the clSTN. One retrograde labelled neuron is found at the transitional border of ilSN towards ilSTN. C5788 contains an injection outside the SN, but the injection touches the middle part of the SNc. The ilSTN contained heavy terminal and fibre labelling. No retrograde-filled ilSTN neurons were found. The clSTN demonstrated sparse fibre and terminal labelling in the upper caudal and middle part of the clSTN (Fig. 29).

6.3.5 Control Injections

Control injections are C5830 for the SNl injections, and C5789 and C5566 for the usual injection tract found above the SN. C5759 and C5754 cover with their injections

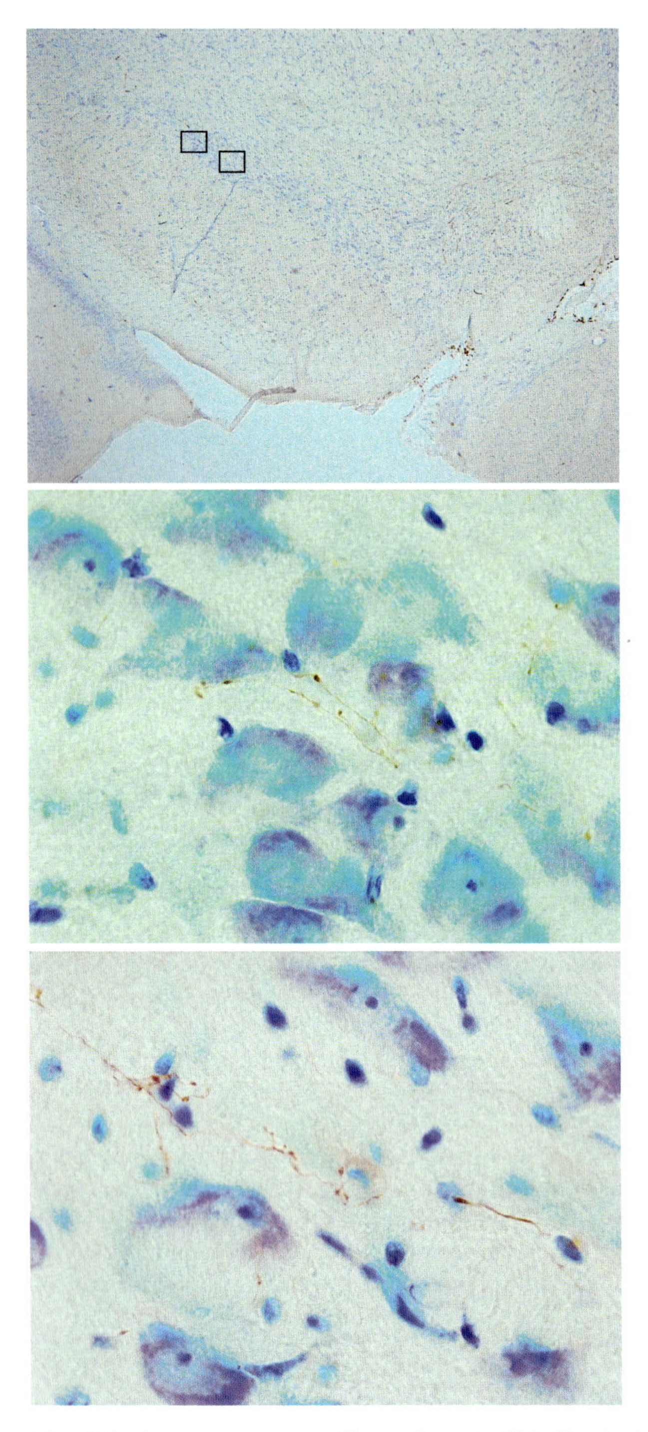

Fig. 29 C5788. The clSTN demonstrates sparse fibre and terminal labelling in the upper caudal and middle part of the clSTN

the whole SN overlaying area. The injections C5560 (caudal ruber) and C5558 (dorsal of ruber and medial of nucleus N III) involve the caudal area of the SN. No labelling was found in the clSTN.

6.4 Discussion

The present study provides data for the existence of a substantial nigrosubthalamic connection in the rat, which also emits a moderate component to the contralateral STN. Thus, two significant nuclei of the basal ganglia—the SN and the STN—are reciprocally strongly interconnected, and this STN-SN-STN loop is involved in the complicated basal ganglia circuitry, since both nuclei display a broad variety of afferent and efferent connections. Generally, the dendrites of the STN projection neurons in the rat, cat and monkey display long, thin dendrites, and in some cases the extent of the dendritic field can almost cover the overall extent of the STN (Iwahori 1978; Romansky 1982; Hammond and Yelnik 1983; Kita et al. 1983; Afsharpour 1985a; Pearson et al. 1985; Romansky and Usunoff 1985, 1987). As noticed also by Hassani et al. (1997), the extent of individual nigrosubthalamic arborizations is considerably smaller than the dimensions of the nucleus. Thus, several different nigrosubthalamic axons probably converge onto a single STN neuron, or, alternatively, a single SN axon might innervate several adjacent STN dendrites.

Many of the efferent connections of SN (to the neostriatum, thalamus, superior colliculus, periaqueductal grey, pedunculopontine tegmental nucleus, red nucleus, mesencephalic nucleus of the trigeminal nerve) are bilateral (Fass and Butcher 1981; Gerfen et al. 1982; Pritzel et al. 1983; Douglas et al. 1987; Ilinsky et al. 1987; Morgan and Huston 1990; Redgrave et al. 1992; Steiner et al. 1992; Lakke et al. 2000; and references therein). Compared to all these connections, the ipsilateral one is considerably larger, and the currently described bilateral nigrosubthalamic projection is no exception. Ipsilaterally the efferent SN axons terminate in large, profuse terminal fields, while contralaterally they terminate in discrete, sharply circumscribed patches. Although the crossed nigrosubthalamic connection is moderate, exactly by its topical distribution, its "point to point" connection is especially evident. The medial SNc projects to the contralateral medial STN, and the lateral SNc also projects mainly to the lateral half of the contralateral STN.

As reviewed in Sect. 5.2, there is already evidence for the DAergic, excitatory nigrosubthalamic connection. Its physiological significance has yet to be unravelled. Unilateral dopamine lesion has been reported to decrease the neuron discharge rate in the contralateral STN, whereas increasing this rate in the ipsilateral STN (Perier et al. 2000). Recently Carr (2002) hypothesized that this pathway might be connected with the rest tremor in Parkinson's disease, e.g. the connections of the STN with the internal pallidum, modified by SN and cortical inputs, allow for the transfer of tremorogenic activity to the thalamus.

The present data also support the suggestion of Ichinohe et al. (2000) for the existence of a moderate projection of parvalbumin containing, presumably GABAergic SN neurons to the STN. Many of the projections of the SNr neurons—to the thalamus, tectum and reticular formation—are built by divergent collaterals of the axon of one and the same SN neuron (Bentivoglio et al. 1979; Beckstead 1983; Parent et al. 1983; Deniau and Chevalier 1992; Yasui et al. 1995; Nishimura et al. 1997). A double-labelling retrograde study might also demonstrate that the non-DAergic afferent connection to the STN is carried out by branching efferent axons of SN, as it is the case with the DAergic nigrosubthalamic tract (Prensa and Parent 2001). The STN-SNr-STN loop consists of descending excitatory component (the glutamatergic subthalamonigral tract), and ascending inhibitory component (the GABAergic nigrosubthalamic tract).

The ultrastructural morphology of the nigrosubthalamic terminal boutons and their participation in the synaptic organization of STN are still unknown. However, some of their features might be predicted. Most probably, the GABAergic nigrosubthalamic boutons share common features with other GABAergic terminals of the pallido-nigral complex: GPE, GPI and SNr, e.g. relatively large boutons containing a pleomorphic synaptic vesicle population, and contacting perikarya and large dendrites by means of symmetric synaptic specializations (Grofová and Rinvik 1974; Romansky et al. 1980a, b; Usunoff et al. 1982a; Kultas-Ilinsky et al. 1983; Williams and Faull 1988; Kultas-Ilinsky and Ilinsky 1990). Therefore, in normal STN ultrastructural material (e.g. Romansky and Usunoff 1987) the GABAergic nigrosubthalamic boutons can hardly be recognized due to the enormous number of pallido-subthalamic terminals (Romansky et al. 1980b; Usunoff et al. 1982a) and this can be reliably examined only by an electron microscopic hodological study. Since the DAergic nigrosubthalamic terminals, at least in part, represent collaterals of the nigrostriatal axons (Gauthier et al. 1999; Prensa and Parent 2001) one might expect that the nigrosubthalamic terminals are relatively small, contain pleomorphic vesicles and form symmetric synapses with various postsynaptic targets, and only rarely form asymmetric synapses with dendritic spines (Hattori et al. 1991; Groves et al. 1994; Hanley and Bolam 1997). Such tyrosine hydroxylase-positive terminals (small size, pleomorphic vesicles and symmetrical axodendritic contacts) were demonstrated in the monkey STN by Smith and Kievel (2000). Although the tyrosine hydroxylase labels noradrenergic terminals too, in all probability the vast majority of these terminals represent nigrosubthalamic terminals.

The STN is a key structure in motor control and should not be regarded only as a relay structure in the so-called indirect pathway by the parallel processing in the basal ganglia circuits (Parent and Hazrati 1995a, b). The STN can still be regarded a "control structure" lying beside the main stream of information processing (cerebral cortex > neostriatum > GPI and SNr > thalamus > cerebral cortex). However, due to its widespread efferent projections (reviewed in Sect. 5.2, this volume), the STN exerts its driving effect on most components of the basal ganglia. Its action is mediated not only by the indirect pathway (cerebral cortex > neostriatum > GPE > STN > GPI and SNr > thalamus > cerebral cortex), but also by a multitude of mono- and polysynaptic projections that ultimately reach the basal ganglia output cells (Parent and Hazrati 1995b).

DAergic medication has been shown to modulate oscillatory activity in the STN and thus may play a role in the pathology of akinesia and rigidity by affecting oscillatory synchronization in the basal ganglia (Allers et al. 2000; Brown et al. 2001; Marsden et al. 2001; Levy et al. 2002). Francois et al. (2000) reported that in the STN of Parkinson's disease patients there is a 65% loss of tyrosine hydroxylase immunoreactive axons (e.g. a 2/3 loss of the STN DAergic innervation) compared with control brains. This significant loss of DAergic innervation might directly affect the activity of the STN neurons, and might participate in the STN hyperactivity.

7 Appendix 1

7.1 Description of the Human Pathology Cases Used in this Study

Series *H3655* concerns a 66-year-old male that suffered from psychotic dementia. A high cervical transverse lesion—due to a fracture of the epistrophic dense, together with a dislocation of the atlas, long before death—was found, resulting in a total cordotomy. The patient died 6 months later of respiratory insufficiency. The cerebrum showed no distortions. The brainstem was stained with Häggqvist technique. Degeneration of the ventrolateral and the dorsal funiculi is massive (for an extensive description of this series see Marani and Schoen 2005).

Series *H5671* concerns a female patient, 56 years old, with a left-side hemiplegia due to a cerebrovascular accident; she died 6 weeks after onset of the start of the accident. After the obduction a softening was found in the right hemisphere, mainly localized in the superior and middle temporal gyri and in the lower part of the central gyri, in the inferior frontal gyrus and in the insula. It extended from there into the lentiform nucleus and corona radiata. The brain stem and spinal cord were stained for Nauta-Gygax and at regular distances frozen Häggqvist sections were made. (For an extensive description see Schoen 1969 and Voogd et al. 1998.)

H5747: The patient, male, 67 years old, died 6 weeks after an operation for otitis media from an otogenic brain abscess in the left temporal lobe. The lesion was well encapsulated in the centre of the posterior part of the temporal lobe, interrupting all afferent and efferent connections of the inferior and middle temporal gyrus. A separate small infarction just dorsal to it effectuated the same for those of the superior temporal convolution. In addition, small lesions in the brain stem interrupted the medial tegmental tract in the medial longitudinal fascicle at the level of the vestibular nuclei, and a lesion in the caudal bulbus was present in the contralateral lemniscus at the same level. The brainstem was frozen and was alternating stained according to Nauta-Gygax, Klüver-Barrera and Häggqvist techniques (see also Marani and Schoen 2005).

Series *H5889*, a Nauta-Gygax, Klüver-Barrera staining, concerns a case of colliquation necrosis in different centres in the rostral brain stem and partially in the medial part of the pes pedunculi. Centres of softening were found in the thalamus (Gebbink 1967) in the ventral part of the rostral mesencephalon, concerning the red nucleus, the neighbouring mesencephalic tegmentum, the substantia nigra and the cerebral peduncle. The lesion extends into the pons, where it continues paramedian from the flm (fasciculus longitudinalis medialis) near to the medial lemniscus. A small lesion was found in the pes pontis and in the area of Wallenberg. The left half of the brain contained a 7-year-old cystic softening tempero-parietally that destroyed the striatum except for its rostroventral part and the capsula interna except for Arnold's tract.

Series *H6348* is also a Nauta-Gygax, Klüver-Barrera series from a 73-year-old male who, after a street accident, showed symptoms of a transverse lesion at C5, together with signs of a severe cerebral contusion. Six weeks after the accident he died, and other than a central cervical cord lesion at C5, a laceration of the right temporal lobe was found. Both ascending spinal ventrolateral and descending corticofugal degeneration were found.

Series *H6368*: In this case the medial part of the *left* half of the midbrain was destroyed by a number of ischaemic lesions presumably 18 days before death. Especially the medial and central tegmental tracts are degenerated and can be followed caudally. Arnold's frontopontine tract is degenerated, whereas Türck's tempero-parietopontine tract is degenerated with a few much smaller connections. The brainstem was frozen and was alternatingly stained according to Nauta-Gygax, Klüver-Barrera and Häggqvist techniques.

8 Appendix 2

8.1 Häggqvist and/or Nauta-Gygax Staining

The material used for this study is summarized in Appendix 1 (description of the pathological cases). Extensive descriptions of the techniques can be found in Usunoff et al. (1997) and Marani and Schoen (2005). Usually the brain and spinal cord were fixed by immersion in 10% formalin within 18 h after death and directly after the autopsy. After several days or weeks the brainstems were dissected out. The tissue blocks and selected spinal cord segments were post fixed for a variable period in Baker's fixative (prior to Häggqvist staining, 6 weeks to 2 months) or in neutral formalin (for Nauta staining 2 weeks to 1 month). For the Häggqvist method the blocks were mordanted in several changes of a 10% solution of potassium dichromate, embedded in paraffin and transversely sectioned at 6 µm. After deparaffination the sections were immersed in 10% phosphomolybdic acid for 30 min, stained in Mann's solution (methyl blue, 0.26%; eosin, water and soluble ethanol, 0.06%), differentiated in ethanol 70%, 96% and 100% and cover slipped. For the Nauta method, 25-µm sections were cut on a Jung freezing microtome and stored in 10% neutral formalin for 1 day to 1 week. The silver impregnation in the protocols of Nauta and Gygax and Nauta and Ebbesson were used with Laidlaw's silver carbonate solution.

8.2 Häggqvist and Klüver-Barrera Staining

Alternate sections were mordanted for 3 days in 5% potassium dichromate at room temperature. After thorough rinsing in tap water the sections were treated with phosphomolybdic acid for 30 min and stained in Mann's solution. Finally the sections were differentiated in ethanol 70% and 96% and mounted from phenol-xylene on glass slides, dried with tissue paper and cover slipped. Additional sections were stained with the Nissl or Klüver-Barrera stain. In most cases some haematoxylin-eosin sections were used to rapidly determine the lesioned area.

8.3 Interpretation of the Staining

In Häggqvist-stained sections axons are coloured blue and the myelin light red. Cell nuclei, cytoplasm, dendrites and glia are stained in different shades of blue. In degenerated fibres the axon has disintegrated or disappeared, and the myelin stains a vivid red, swells and becomes vacuolated. Häggqvist's stain was originally developed by Alzheimer as a glial stain. The gliosis and the compound granular cells that predominate in chronic degeneration therefore can be studied at advantage. In the silver-impregnated sections, increased argyrophilia, fragmentation and vacuolization are considered as signs of degeneration of the axon. The presence of a meshwork of fine, degenerated axons in a nucleus is taken as a sign of termination and called "preterminal" to distinguish it from bouton degeneration, which cannot been observed. Different kinds of artefacts were frequently encountered. Dust-like precipitates, "myelin cuffs" and irregular, fusiform enlargements of apparently normal fibres generally could be distinguished from true axonal degeneration.

References

Afsharpour S (1985a) Light microscopic analysis of Golgi-impregnated rat subthalamic neurons. J Comp Neurol 236:1–13

Afsharpour S (1985b) Topographical projections of the cerebral cortex to the subthalamic nucleus. J Comp Neurol 236:14–28

Albin RL, Aldridge JW, Young AB, Gilman S (1989a) Feline subthalamic nucleus neurons contain glutamate-like but not GABA-like or glycine-like immunoreactivity. Brain Res 491:185–188

Albin RL, Young AB, Penny JB (1989b) The functional anatomy of basal ganglia disorders. Trends Neurosci 12:366–375

Allers KA, Kreiss DS, Walters JR (2000) Multisecond oscillations in the subthalamic nucleus: effects of apomorphine and dopamine cell lesion. Synapse 38:38–50

Altman J (1963) Regional utilization of leucine-H3 by normal rat brain: microdensitometric evaluation of autoradiograms. J Histochem Cytochem 11:741–751

Alvarez L, Macias R, Guridi J, Lopez G, Alvarez E, Maragoto C, Teijeiro J, Torres A, Pavon N, Rodriguez-Oroz MC, Ochoa L, Hetherington H, Juncos J, DeLong MR, Obeso JA (2001) Dorsal subthalamotomy for Parkinson's disease. Mov Disord 16:72–78

Armstrong DM, Miller RJ, Beaudet A, Pickel VM (1984) Enkephalin-like immunoreactivity in rat area postrema: ultrastructural localization and coexistence with serotonin. Brain Res 310:269–278

Arvidsson U, Ulfhake B, Cullheim S, Bergstrand A, Theodorson E, Hokfelt T (1991) Distribution of 125I-galanin binding sites, immunoreactive galanin, and its coexistence with 5-hydroxytryptamine in the cat spinal cord: biochemical, histochemical, and experimental studies at the light and electron microscopic level. J Comp Neurol 308:115–138

Ashby P, Kim YJ, Kumar R, Lang AE, Lozano AM (1999) Neurophysiological effects of stimulation through electrodes in the human subthalamic nucleus. Brain 122:1919–1931

Ashby P, Paradiso G, Saint-Cyr JA, Chen R, Lang AE, Lozano AM (2001) Potentials recorded at the scalp by stimulation near human subthalamic nucleus. Clin Neurophysiol 112:431–437

Auer J (1956) Terminal degeneration in the diencephalon after ablation of frontal cortex in the cat. J Anat 90:30–41

Augood SJ, Waldvogel HJ, Munkle MC, Faull RLM, Emson PC (1999) Localization of calcium-binding proteins and GABA transporter (GAT-1) messenger RNA in the human subthalamic nucleus. Neuroscience 88:521–534

Augood SJ, Hollingsworth ZR, Standaert DG, Emson PC, Penney Jr JB (2000) Localization of dopaminergic markers in the human subthalamic nucleus. J Comp Neurol 421:247–255

Awad H, Hubert GW, Smith Y, Levey AI, Conn PJ (2000) Activation of metabotropic glutamate receptor 5 has direct excitatory effects and potentiates NMDA receptor currents in neurons of the subthalamic nucleus. J Neurosci 20:7871–7879

Baker KB, Montgomery EB Jr, Rezai AR, Burgess R, Luder HO (2002) Subthalamic nucleus deep brain stimulation evoked potentials: physiological and therapeutic implications. Mov Disord 17:969–983

Barbè A (1938) Recherches sur l'embryologie du systéme nerveux central de l'homme. Masson et cie, Paris

Battistin L, Scarlato G, Caraceni T, Ruggieri S (1996) Parkinson's disease. Advances in nuerology, vol 69 Lippincott-Raven, Philadelphia

Baufreton J, Zhu ZT, Garret M, Bioulac B, Johnson SW, Taupignon AI (2005) Dopamine receptors set the pattern of activity generated in subthalamic neurons. FASEB J 19:1771–1777

Beaudet A, Descarries L (1981) The fine structure of central serotonin neurons. J Physiol 77:193–203

Beckstead RM (1983) Long collateral branches of substantia nigra pars reticulata axons to thalamus, superior colliculus and reticular formation in monkey and cat. Multiple retrograde neuronal labeling with fluorescent dyes. Neuroscience 10:767–779

Bell K, Churchill L, Kalivas PW (1995) GABAergic projection from the ventral pallidum and globus pallidus to the subthalamic nucleus. Synapse 20:10–18

Benabid AL (2003) Deep brain stimulation for Parkinson's disease. Curr Opin Neurobiol 13:696–706

Benabid AL, Pollak P, Louveau A, Henry S, de Rougemont J (1989) Combined (thalamotomy and stimulation) stereotactic surgery of the VIM thalamic nucleus for bilateral Parkinson disease. Appl Neurophysiol 50:344–346

Benabid AL, Krack P, Benazzouz A, Limousin P, Koudsie A, Pollak P (2000) Deep brain stimulation of the subthalamic nucleus for Parkinson's disease: methodologic aspects and clinical criteria. Neurology 55:S40–S44

Benazzouz A, Gao D, Ni Z, Benabid AL (2000a) High frequency stimulation of the STN influences the activity of dopamine neurons in the rat. Neuroreport 11:1593–1596

Benazzouz A, Gao DM, Ni ZG, Piallat B, Bouali-Benazzouz R, Benabid AL (2000b) Effect of high-frequency stimulation of the subthalamic nucleus on the neuronal activities of the substantia nigra pars reticulate and ventrolateral nucleus of the thalamus in the rat. Neuroscience 99:289–295

Benazzouz A, Breit S, Koudsie A, Pollak P, Krack P, Benabid A (2002) Intraoperative microrecordings of the subthalamic nucleus in parkinson's disease. Mov Disord 17:S145–S149

Bentivoglio M, van der Kooy D, Kuypers HG (1979) The organization of the efferent projections of the substantia nigra in the rat. A retrograde fluorescent double labeling study. Brain Res 174:1–17

Berendse HW, Groenewegen HJ (1991) The connections of the medial part of the subthalamic nucleus in rat: evidence for a parallel organization. In: Bernardi G, et al (eds) The basal ganglia III. Plenum, New York, pp 89–98

Bergman H, Wichmann T, DeLong MR (1990) Reversal of experimental parkinsonism by lesions of the subthalamic nucleus. Science 249:1436–1438

Bergman H, Wichmann T, Karmon B, DeLong MR (1994) The primate subthalamic nucleus. II. Neuronal activity in the MPTP model of parkinsonism. J Neurophysiol 72:507–520

Beurrier C, Congar P, Bioulac B, Hammond C (1999) Subthalamic nucleus neurons switch from single spike activity to burst-firing mode. J Neurosci 19:599–609

Bevan MD, Bolam JP (1995) Cholinergic, GABAergic, and glutamate-enriched inputs from the mesopontine tegmentum to the subthalamic nucleus in the rat. J Neurosci 15:7105–7120

Bevan MD, Francis CM, Bolam JP (1995) The glutamate-enriched cortical and thalamic input to neurons in the subthalamic nucleus of the rat: convergence with GABA-positive terminals. J Comp Neurol 361:491–511

Bevan MD, Magill P, Terman D, Bolam J, Wilson C (2002) Move to the rhythm: oscillations in the subthalamic nucleus-external globus pallidus network. Trends Neurosci 25:525

Billinton A, Ige AO, Wise A, White JH, Disney GH, Marshall FH, Waldvogel HJ, Faull R, Emson PC (2000) GABAB receptor heterodimer-component localisation in human brain. Brain Res Mol Brain Res 77:111–124

Bischoff S, Leonhard S, Reymann N, Schuler V, Shigemoto R, Kaupmann K, Bettler B (1999) Spatial distribution of GABABR1 receptor mRNA and binding sites in the rat brain. J Comp Neurol 412:1–16

Björklund A, Lindvall O (1984) Dopamine-containing systems in the CNS. In: Bjorklund A, Hokfelt T (eds) Handbook of chemical neuroanatomy, vol 2, part I. Classical transmitters in the CNS. Elsevier, Amsterdam, pp 55–122

Blandini F, Nappi G, Tassoreli C, Martignoni E (2000) Functional changes of the basal ganglia circuitry in Parkinson's disease. Prog Neurobiol 62:63–88

Bobillier P, Seguin S, Petitjean F, Salvert D, Touret M, Jouvet M (1976) The raphe nuclei of the cat brain stem: a topographical atlas of their efferent projections as revealed by autoradiography. Brain Res 113:449–486

Bouthenet ML, Martres MP, Sales N, Schwartz JC (1987) A detailed mapping of dopamine D2 receptors in the rat central nervous system by autoradiography with (125I) iodosulpuride. Neuroscience 20:117–155

Bouthenet ML, Soul E, Martres MP, Sokoloff P, Giros B, Schwartz JC (1991) Localization of dopamine D3 receptor mRNA in the rat brain using in situ histochemistry: comparison with dopamine D2 receptor mRNA. Brain Res 564:203–219

Boyes J, Bolam JP (2007) Localization of GABA receptors in the basal ganglia. Prog Brain Res 160:229–243

Boyson SJ, McGonigle P, Molinoff PB (1986) Quantitative autoradiographic localization of the D1 and D2 subtypes of dopamine receptors in rat brain. J Neurosci 6:3177–3188

Brodal A (1981) Neurological anatomy in relation to clinical medicine, 3rd edn. Oxford University Press, Oxford

Broere G (1971) Corticofugal fibers in some mammals. Thesis. Leiden University

Brotchie JM, Crossman AR (1991) D-[^{3}H]aspartate and [14C]GABA uptake in the basal ganglia of rats following lesions in the subthalamic region suggest a role for excitatory amino acid but not GABA-mediated transmission in subthalamic nucleus efferents. Exp Neurol 113:171–181

Brown LL, Makman MH, Wolfson LI, Dworkin B, Wagner C, Katzman R (1979) A direct role of dopamine in the rat subthalamic nucleus and an adjacent intrapeduncular area. Science 206:1416–1418

Brown P, Oliviero A, Mazzone P, Insola A, Tonali P, Di Lazzaro V (2001) Dopamine dependency of oscillations between subthalamic nucleus and pallidum in Parkinson's disease. J Neurosci 21:1033–1038

Bruinvels AT, Palacios JM, Hoyer D (1993) Autoradiographic characterisation and localisation of 5-HT1D compared to 5-HT1B binding sites in rat brain. Naunyn Schmiedebergs Arch Pharmacol 347:569–582

Buijs RM, Geffard M, Pool CW, Hoorneman EM (1984) The dopaminergic innervation of the supraoptic and paraventricular nucleus. A light and electron microscopical study. Brain Res 323:65–72

Bulfone A, Puelles L, Porteus MH, Frohman MA, Martin GR, Rubenstein JLR (1993) Spatially restricted expression of Dlx-1, Dlx-2 (Tes-1), Gbx-2 and Wnt-3 in the embryonic day 12. 5 mouse forebrain defines potential transverse and longitudinal segmental boundaries. J Neurosci 13:3155–3172

Cabana T, Martin GF (1986) Development of projections from somatic motor sensory areas of neocortex to the diencephalon and brainstem in the North American opossum. J Comp Neurol 251:506–516

Campbell GA, Eckhardt MJ, Weight FF (1985) Dopaminergic mechanisms in subthalamic nucleus of rat: analysis using horseradish peroxidase and microiontophoresis. Brain Res 333:261–270

Canteras NS, Shammah-Lagnado SJ, Silva BA, Ricardo JA (1988) Somatosensory inputs to the subthalamic nucleus: a combined retrograde and anterograde horseradish peroxidase study in the rat. Brain Res 458:53–64

Canteras NS, Shammah-Lagnado SJ, Silva BA, Ricardo JA (1990) Afferent connections of the subthalamic nucleus; a combined retrograde and anterograde horse radish peroxidase study in the rat. Brain Res 513:43–59

Carden WB, Datskovskaia A, Guido W, Godwin DW, Bickford ME (2000) Development of cholinergic, nitrergic, and GABAergic innervation of the cat dorsal lateral geniculate nucleus. J Comp Neurol 418:65–80

Carpenter MB, Carpenter CS (1951) Analysis of somatotopic relations of the corpus Luysi in man and monkey. J Comp Neurol 95:349–370

Carpenter MB, Strominger NL (1967) Efferent fibers of the subthalamic nucleus in the monkey. A comparison of the efferent projections of the subthalamic nucleus, substantia nigra and globus pallidus. Am J Anat 121:41–72

Carpenter MB, Whittier JR, Mettler FA (1950) Analysis of choreoid hyperkinesia in the rhesus monkey. J Comp Neurol 92:293–312

Carpenter MB, Fraser RA, Shriver JE (1968) The organization of pallidosubthalamic fibers in the monkey. Brain Res 11:522–559

Carpenter MB, Baton RR III, Carleton SC, Keller JT (1981) Interconnections and organization of pallidal and subthalamic nucleus neurons in the monkey. J Comp Neurol 197:579–603

Carr J (2002) Tremor in Parkinson's disease. Parkinsonism Relat Disord 8:223–234

Carrive P (1993) The periaquaductal gray and defensive behavior: functional representation and neuronal organization. Behav Brain Res 58:27–47

Castle M, Aymerich MS, Sanchez-Escobar C, Gonzalo N, Obeso JA, Lanciego JL (2005) Thalamic innervation of the direct and indirect basal ganglia pathways in the rat: ipsi- and contralateral projections. J Comp Neurol 483:143–153

Cha CI, Sohn SG, Chung YH, Shin C, Baik SH (2000) Region specific changes of NOS-IR cells in the basal ganglia of the aged rat. Brain Res 854:239–244

Chan-Palay V (1977) Cerebellar dentate nucleus: organization, cytology and transmitters. Springer-Verlag, Berlin Heidelberg New York

Chan-Palay V (1982) Serotonin neurons and their axons in the raphe dorsalis of rat and rhesus monkey: demonstration by high resolution radioautography with 3H serotonin. In: Chan-Palay V, Palay SL (eds) Cytochemical methods in neuroanatomy. Alan Liss, New York, pp 357–386

Chang HT, Kita H, Kitai ST (1983) The fine structure of the rat subthalamic nucleus: an electron microscopic study. J Comp Neurol 221:113–123

Chang HT, Kita H, Kitai ST (1984) The ultrastructural morphology of the subthalamo-nigral axon terminals intracellularly labeled with horseradish peroxidase. Brain Res 299:182–185

Charara A, Heilman C, Levey Al, Smith Y (2000) Pre and postsynaptic localization of GABAB receptor in basal ganglia in monkeys. Neuroscience 95:127–140

Chen JF, Xu K, Petzer JP, Staal R, Xu YH, Bellstein M, Sonsalla PK, et al (2001) Neuroprotection by caffeine and A2A adenosine receptor inactivation in a model of Parkinson's disease. J Neurosci 21:143–149

Chen L, Boyes J, Yung WH, Bolam JP (2004) Subcellular localization of GABA B receptor subunits in rat globus pallidus. J Comp Neurol 474:340–352

Ciliax BJ, Greenamyre T, Levey AL (1997) Functional biochemistry and molecular neuropharmacology of the basal ganglia and motor systems. In: Watts RL (ed) Movement disorders: neurological principles and practice. Mc Graw-Hill, New York, pp 99–116

Cimino M, Marini P, Fornasari D, Cattabeni F, Clementi F (1992) Distribution of nicotinic receptors in cynomolgus monkey brain and ganglia: localization of α3 subunit mRNA, α-bungarotoxin and nicotine binding sites. Neuroscience 51:77–86

Clarke NP, Bolam JP (1998) Distribution of glutamate receptor subunits at neurochemically characterized synapses in the entopeduncular nucleus and subthalamic nucleus of the rat. J Comp Neurol 397:403–420

Clements JR, Grant S (1990) Glutamate-like immunoreactivity in neurons of the laterodorsal tegmental and pedunculopontine nuclei in the rat. Neurosci Lett 120:70–73

Clements JR, Toth DD, Highfield DA, Grant S (1991) Glutamate-like immunoreactivity is present within cholinergic neurons of the laterodorsal tegmental and pedunculopontine nuclei. Adv Exp Med Biol 295:127–142

Coizet V, McHaffie JG, May P, Stanford TR, Jiang H, et al (2007) The tecto-subthalamic projection: a source of short-latency visual input to the subthalamic nucleus. http://www.ibags.info. Cited 8 Feb 2008

Cole RI, Lechner SM, Williams ME, Prodanovich P, Bleicher L, Varney MA, Guibao G (2005) Differential distribution of voltage-gated alpha-2 delta (α2δ) subunit mRNA-containing cells in the rat central nervous system and the dorsal root ganglia. J Comp Neurol 491:246–269

Cossette M, Levesque M, Parent A (1999) Extrastriatal dopaminergic innervation of human basal ganglia. Neurosci Res 34:51–54

Cote PY, Sadikot AF, Parent A (1991) Complementary distribution of calbindin D-28k and parvalbumin in the basal forebrain and midbrain of the squirrel monkey. Eur J Neurosci 3:1316–1329

Cragg SJ, Baufreton J, Xue Y, Bolam JP, Bevan MD (2004) Synaptic release of dopamine in the subthalamic nucleus. Eur J Neurosci 20:1788–1802

Crossman AR (1987) Primate models of dyskinesia: the experimental approach to the study of basal ganglia-related involuntary movement disorders. Neuroscience 21:1–40

De Vito JL, Smith OA Jr (1964) Subcortical projections of the prefrontal lobe of the monkey. J Comp Neurol 123:413–424

Dejerine J (1901) Anatomie des centres nerveux. Tombe II, Rueff et cie Paris

Delfs JM, Kong H, Mestek A, Chen Y, Yu L, et al (1994) Expression of mu opioid receptor mRNA in rat brain: an in situ hybridization study at the single cell level. J Comp Neurol 345:46–68

den Dunnen WF, Staal MJ (2005) Anatomical alterations of the subthalamic nucleus in relation to age: a postmortem study. Mov Disord 20:893–898

Deniau JM, Chevalier G (1992) The lamellar organization of the rat substantia nigra pars reticulata: distribution of projection neurons. Neuroscience 46:361–377

Deutch AY, Holliday j, Roth RH, Chun LLY, Hawrot E (1987) Immunohistochemical localization of a neuronal nicotinic acetylcholine receptor in mammalian brain. Proc Natl Acad Sci USA 84:8697–8701

DiFiglia M, Aronin N (1982) Ultrastructural features of immunoreactive somatostatin neurons in the rat caudate nucleus. J Neurosci 2:1267–1274

Dormont D, Kenneth GR, Tandé D, Parain K, et al (2004) Is the subthalamic nucleus hypointense on T2-weighted images? A correlation study using imaging and stereotactic atlas data. AJNR Am J Neuroradiol 25:1516–1523

Douglas R, Kellaway L, Mintz M, van Wageningen G (1987) The crossed nigrostriatal projection decussates in the ventral tegmental decussation. Brain Res 418:111–121
Durkin MM, Grunwaldsen CA, Borowsky B, Jones KA, Branchet TA (1999) An in situ hybridization study of the distribution of the GABA B2 protein μ RNA in the rat CNS. Mol Brain Res 71:185–200
Eberle-Wang K, Mikeladze Z, Uryu K, Chesselet MF (1997) Pattern of expression of the serotonin 2C receptor messenger RNA in the basal ganglia of adult rats. J Comp Neurol 384:233–247
Eve DJ, Nisbet AP, Kingsburry AE, Hewson EL, Daniel SE, Lees AJ, Marsden CD, Forster OJ (1998) Basal ganglia neuronal nitric oxide synthase mRNA expression in Parkinson's disease. Brain Res Mol Brain Res 63:62–71
Fass B, Butcher LL (1981) Evidence for a crossed nigrostriatal pathway in rats. Neurosci Lett 22:109–113
Faull RLM, Carman JB (1978) The cerebellofugal projections in the brachium conjunctivum of the rat. I. The contralateral ascending pathway. J Comp Neurol 178:495–517
Feger J, Hammond C, Rouzaire-Dubois B (1979) Pharmacological properties of acetylcholine-induced excitation of subthalamic nucleus neurons. Br J Pharmacol 65:511–515
Feger J, Hassani OK, Mouroux M (1997) The subthalamic nucleus and its connections. New electrophysiological and pharmacological data. Adv Neurol 74:31–43
Feirabend HKP, Choufoer H, Ploeger S, Holsheimer J, van Gool JD (2002) Morphometry of human superficial dorsal and dorsolateral column fibres: significance to spinal cord stimulation. Brain 125:1137–1149
Fischer O (1911) Zur Frage der anatomischen Grundlage der Athetose double und der posthemiplegischen Bewegungstoerung ueberhaupt. Z Neurol 7:463
Flechsig P (1876) Die Leitungsbahnen im Gehirn und Rueckenmark des Menschen auf Grund entwicklungsgeschichtlicher Untersuchungen. Leipzig
Floran B, Floran L, Erlij D, Aceves J (2004) Dopamine D4 receptors inhibit depolarization-induced [^{3}H]GABA release in the rat subthalamic nucleus. Eur J Pharmacol 498:97–102
Flores G, Hernandez S, Rosales MG, Sierra A, Martines-Fong D, Flores Hernandez J (1996) M3 muscarine receptors mediate cholinergic excitation of the spontaneous activity of the subthalamic neurons in the rat. Neurosci Lett 203:203–206
Flores G, Liang JJ, Sierra A, Martinezfong D, Quiron R, Aceves J, Srivastava LK (1999) Expression of dopamine receptors in the subthalamic nucleus of the rat: characterization using reverse transcriptase polymerase chain reaction and autoradiography. Neuroscience 91:549–556
Florin S, Meunier JC, Costentin J (2000) Autoradiographic localization of [^{3}H] nociceptin binding sites in the rat brain. Brain Res 880:11–16
Forel A (1872) Beitrage zur Kenntnis des Thalamus opticus und der ihn umgebenden Geblde bei Sãugetieren. Sitzbeitrag d.K.Akad.d.Wissensch.zu Wien Math-Nat CL.BD 66, Abtr 3 S.25
Forel A (1877) Untersuchungen ueber die Haubenregion und ihre oberen Verknuepfung im Gehirne des Menschen und einiger Säugetiere, mit Beiträgen zu den Methoden der Gehirnuntersuchung. Archiv f. Psychiatrie Band VII
Fortin M, Parent A (1996) Calretinin as a marker of specific neuronal subsets in primate substantia nigra and subthalamic nucleus. Brain Res 708:201–204
Francois C, Savy C, Jan C, Tande D, Hirsch EC, Yelnik J (2000) Dopaminergic innervation of the subthalamic nucleus in the normal state, in MPTP-treated monkeys, and in Parkinson's disease patients. J Comp Neurol 425:121–129
Freiman J, Szabo B (2005) Cannabinoids depress excitatory neurotransmission between the subthalamic nucleus and globus pallidus. Neuroscience 133:305–313

Friede RL (1966) Topographic brain chemistry. Academic Press, New York
Fritschy JM, Möhler H (1995) GABAA-receptor heterogeneity in the adult rat brain: differential regional and cellular distribution of seven major subunits. J Comp Neurol 359:154–194
Fujimoto K, Kita H (1993) Response characteristics of subthalamic neurons to the stimulation of the sensorimotor cortex in the rat. Brain Res 609:185–192
Fujiyama F, Fritschy JM, Stephenson FA, Bolam JP (2003) Synaptic localization of GABA-A receptor units in the basal ganglia of the rat. In: Graybiel AM, DeLong MR, Kitai ST (eds) Basal ganglia VI. Plenum Press, New York, pp 631–640
Füssenich MS (1967) Vergleichend anatomische Studien ueber den Nucleus subthalamicus (Corpus Luysi) bei Primaten. Diss C&O Vogt Institut fuer Hirnforschung, Neustadt/ Albert-Ludwigs Universitaet Frieburg
Fuster JM, Uyeda AA (1962) Facilitation of tachiscopic performance by stimulation of midbrain tegmental point in the monkey. Exp Neurol 6:384–406
Galvan A, Charara A, Pare JF, Levey AJ, Smith Y (2004) Differential subcellular and subsynaptic distribution of GABA-A and GABA-B receptors in monkey subthalamic nucleus. Neuroscience 127:709–721
Garcia-Rill E (1991) The pedunculopontine nucleus. Prog Neurobiol 36:363–389
Gauthier J, Parent M, Levesque M, Parent A (1999) The axonal arborization of single neostriatal neurons in rats. Brain Res 834:228–232
Gebbink TB (1967) Structure and connections of the basal ganglia in man. Van Gorcum, Assen
Gebbink TB (1969) Some remarks on the ansa lenticularis in man. Psychiatr Neurol Neurochir 72:37–43
Gerfen CR, Wilson CJ (1996) The basal ganglia. In: Swanson LW, Bjorklund A, Hokfelt T (eds) Handbook of chemical neuroanatomy integrated systems in the CNS, part III. Elsevier, Amsterdam, pp 371–468
Gerfen CR, Staines WA, Arbuthnott GW, Fibiger HC (1982) Crossed connections of the substantia nigra in the rat. J Comp Neurol 207:283–303
Geula C, Schatz CR, Mesulam MM (1993) Differential localization of NADPH-diaphorase and calbindin-D28k within the cholinergic neurons of the basal forebrain, striatum and brainstem in the rat, monkey, baboon and human. Neuroscience 54:461–476
Glass M, Dragunow M, Faull RL (1997) Cannabinoid receptors in the human brain: a detailed and quantitative autoradiographic study in fetal, neonatal and adult human brain. Neuroscience 77:299–318
Glees P, Wall PD (1946) Fiber connections of the subthalamic region and the centromedian nucleus of the thalamus. Brain 69:195–208
Goss JR, Morgan DG (1995) Enhanced glial fibrillary acidic protein RNA response to fornix transaction in aged mice. J Neurochem 64:1351–1360
Granata AR, Kitai ST (1989) Intracellular study of nucleus parabrachialis and nucleus tractus solitarii interconnections. Brain Res 492:281–292
Graybiel AM (1977) Direct and indirect preoculomotor pathways of the brainstem: an autoradiographic study of the pontine reticular formation in cat. J Comp Neurol 175:37–78
Greidenberg N (1882) Vier Faelle von Athetose. Petersb med Wschr 23
Greiff (1883) Zur Lokalisation der Hemichorea. Arch Psych 14:598
Groenewegen HJ, Berendse HW (1990) Connections of the subthalamic nucleus with ventral striatopallidal parts of the basal ganglia in the rat. J Comp Neurol 294:607–622
Grofová I (1965) Frontal cortical projections to the midbrain tegmentum in the cat. Folia Morphol (Praha) 13:305–315
Grofová I (1969) Experimental demonstration of a topical arrangement of the pallidosubthalamic fibers in the cat. Psychiatr Neurol Neurochir 72:53–59

Grofová I, Rinvik E (1971) Effect of reserpine on large dense-core vesicles in the boutons in the cat's substantia nigra. Neurobiology 1:5–16

Grofová I, Rinvik E (1974) Cortical and pallidal projections to the nucleus ventralis lateralis thalami. Electron microscopical studies in the cat. Anat Embryol (Berl) 146:113–132

Grofová I, Zhou M (1998) Nigral innervation of cholinergic and glutamatergic cells in the rat mesopontine tegmentum: light and electron microscopic anterograde tracing and immunohistochemical studies. J Comp Neurol 395:359–379

Grosche J, Matyash V, Moller T, Verkhratsky A, Reichenbach A, Kettenmann H (1999) Microdomains for neuron-glia interaction: parallel fiber signaling to Bergmann glial cells. Nat Neurosci 2:139–143

Gross RE, Lozano AM (2000) Advances in neurostimulation for movement disorders. Neurol Res 22:247–258

Groves PM, Wilson CJ (1980) Monoaminergic presynaptic axons and dendrites in locus coeruleus as seen in reconstructions of serial sections. J Comp Neurol 193:853–862

Groves PM, Linder JC, Young SJ (1994) 5-Hydroxydopamine-labeled dopaminergic axons: three-dimensional reconstructions of axons, synapses and postsynaptic targets in rat neostriatum. Neuroscience 58:593–604

Guridi J, Obeso JA (2001) The subthalamic nucleus, hemiballismus and Parkinson's disease: reappraisal of a neurosurgical dogma. Brain 124:5–19

Guridi J, Luquin MR, Herrero MT, Obeso JA (1993) The subthalamic nucleus: a possible target for stereotaxic surgery in Parkinson's disease. Mov Disord 8:421–429

Haartsen AB (1962) Cortical projections to mesencephalon, pons, medulla oblongata and spinal cord. Thesis. Leiden University

Hallervorden J (1957) Die Torsionsdystonie. Der Hemiballismus In: Lubarsch O, Henke I (eds) Handbuch der speziellen pathologischen Anatomie und Histologie 13 Part 1A. Springer, Berlin Heidelberg New York, pp 1957

Hamani C, Saint-Cyr JA, Fraser J, Kaplitt M, Lozano AM (2004) The subthalamic nucleus in the context of movement disorders. Brain 127:4–20

Hamilton BI, Skultety FM (1970) Efferent connections of the periaqueductal gray matter in cat. J Comp Neurol 139:105–114

Hammond C, Rouzaire-Dubois B, Feger J, Jackson A, Crossman AR (1983a) Anatomical and electrophysiological studies on the reciprocal projections between the subthalamic nucleus and nucleus tegmenti pedunculopontinus in the rat. Neuroscience 9:41–52

Hammond C, Shibazaki T, Rouzaire-Dubois B (1983b) Branched output neurons of the rat subthalamic nucleus: electrophysiological study of the synaptic effects on identified cells in the two main target nuclei, the entopeduncular nucleus and the substantia nigra. Neuroscience 9:511–520

Hammond C, Yelnik J (1983c) Intracellular labelling of rat subthalamic neurones with horseradish peroxidase: computer analysis of dendrites and characterization of axon arborization. Neuroscience 8:781–790

Hanajima R, Ashby P, Lozano AM, Lang EL, Chen R (2004) Single pulse stimulation of the human subthalamic nucleus facilitates the motor cortex at short intervals. J Neurophysiol 92:1937–1943

Hanley JJ, Bolam JP (1997) Synaptology of the nigrostriatal projection in relation to the compartmental organization of the neostriatum in the rat. Neuroscience 81:353–370

Hardman GD, Henderson JM, Finkelstein DI, Horne MK, Paxinos G, Halliday GM (2002) Comparison of the basal ganglia in rats, marmosets, macaques, baboons, and humans: volume and neuronal number for the output, internal relay and striatal modulating nuclei. J Comp Neurol 445:238–255

Hartmann-Von Monakow K, Akert K, Künzle H (1978) Projections of the precentral motor cortex and other cortical areas of the frontal lobe to the subthalamic nucleus in the monkey. Exp Brain Res 33:395–403

Hassani OK, Mouroux M, Feger J (1996) Increased subthalamic neuronal activity after nigral dopaminergic lesion independent of disinhibition via the globus pallidus. Neuroscience 72:105–115

Hassani OK, Francois C, Yelnik J, Feger J (1997) Evidence for a dopaminergic innervation of the subthalamic nucleus in the rat. Brain Res 749:88–94

Hassler R, Bak IJ, Kim JS (1970) Unterschiedliche Entleerung der Speicherorte fur Noradrenalin, Dopamin und Serotonin als Wirkungsprinzip des Oxypertins. Nervenarzt 41:105–118

Hassler R, Usunoff KG, Romansky KV, Christ JF (1982) Electron microscopy of the subthalamic nucleus in the baboon. I. Synaptic organization of the subthalamic nucleus in the baboon. J Hirnforsch 23:597–611

Hattori T, Takada M, Moriizumi T, van der Kooy D (1991) Single dopaminergic nigrostriatal neurons form two chemically distinct synaptic types: possible transmitter segregation within neurons. J Comp Neurol 309:391–401

Hedreen JC (1999) Tyrosine hydroxylase-immunoreactive elements in the human globus pallidus and subthalamic nucleus. J Comp Neurol 409:400–410

Henle J (1879) Handbuch der Nervenlehre des Menschen. Handbuch der systematischen Anatomie des Menschen. Braunschweig

Herkenham M, Lynn AB, De Costa BR, Richfield EK (1991) Neuronal localization of cannabinoid receptors in the basal ganglia of the rat. Brain Res 547:267–274

Hess WR (1949) Das Zwischenhirn. Syndrome, Lokalisationen, Funktionen. Benno Schwabe, Basel, pp 1–187

Hirasawa K, Okano S, Kamio S (1938) Beitrag zur Kenntnis ueber die corticalen extrapyramidalen Fasern aus der Area temp. Z Mikrosk Anat Forsch 44[Suppl 22]:74–84

Hironishi M, Ueyama E, Senba E (1999) Systematic expression of immediate early genes and intensive astrocyte activation induced by intrastriatal ferrous iron injection. Brain Res 828:145–153

Hirsch EC, Perier C, Orieux G, Francois C, Feger J, Yelnik J, Vila M, Levy R, Tolosa ES, Marin C, Trinidad Herrero M, Obeso JA, Agid Y (2000) Metabolic effects of nigrostriatal denervation in basal ganglia. Trends Neurosci 23:S78–S85

Hirsch EC, Breidert T, Rousselet E, Hunot S, Hartmann A, Michel PP (2003) The role of glial reaction and inflammation in Parkinson's disease. Ann NY Acad Sci 991:214–228

Hokfelt T, Skirboll L, Rehfeld JF, Goldstein M, Markey K, Dann O (1980) A subpopulation of mesencephalic dopamine neurons projecting to limbic areas contains a cholecystokinin-like peptide: evidence from immunohistochemistry combined with retrograde tracing. Neuroscience 5:2093–2124

Hollerman JR, Grace AA (1992) Subthalamic nucleus cell firing in the 6-OHDA treated rat: basal activity and response to haloperidol. Brain Res 590:291–299

Hontanilla B, Parent A, Gimenez-Amaya JM (1997) Parvalbumin and calbindin D-28k in the entopeduncular nucleus, subthalamic nucleus, and substantia nigra of the rat as revealed by double-immunohistochemical methods. Synapse 25:359–367

Hontanilla B, Parent A, de las Heras S, Gimenez-Amaya JM (1998) Distribution of calbindin D-28k and parvalbumin neurons and fibers in the rat basal ganglia. Brain Res Bull 47:107–116

Hoogstraten MC, Lakke JPWF, Zwarts MJ (1986) Bilateral ballism: a rare syndrome. J Neurol 233:25–29

Houeto JL, Mesnage V, Mallet L, Pillon B, Gargiulo M, du Moncel ST, Bonnet AM, Pidoux B, Dormont D, Cornu P, Agid Y (2002) Behavioural disorders, Parkinson's disease and subthalamic stimulation. J Neurol Neurosurg Psychiatr 72:701–707

Huang Z, Kwong H, Gibb A (2007) NMDA receptor subtypes in neurons of the rat subthalamic nucleus. Life Sci Proc Life Sci PC 398

Hurd YL, Suzuki M, Sedvall GC (2001) D1 and D2 dopamine receptor mRNA expression in whole hemisphere sections of the human brain. J Chem Neuroanat 22:127–137

Hurley MJ, Mash DC, Jenner P (2004) Expression of cannabinoid CB1 receptor m-RNA in basal ganglia of normal and parkisonian human brain. J Neural Transm 110:1279–1288

Ichinohe N, Teng B, Kitai ST (2000) Morphological study of the tegmental pedunculopontine nucleus, substantia nigra and subthalamic nucleus, and their interconnections in rat organotypic culture. Anat Embryol (Berl) 201:435–453

Ilinsky IA, Kultas-Ilinsky K, Rosina A, Hassy M (1987) Quantitative evaluation of crossed and uncrossed projections from basal ganglia and cerebellum to the cat thalamus. Neuroscience 21:207–227

Inglis WL, Winn P (1995) The pedunculopontine tegmental nucleus: where the striatum meets the reticular formation. Prog Neurobiol 47:1–29

Iwahori N (1978) A Golgi study on the subthalamic nucleus of the cat. J Comp Neurol 182:383–397

Jackson A, Crossman AR (1981) Subthalamic nucleus efferent projection to the cerebral cortex. Neuroscience 6:2367–2377

Jackson A, Crossman AR (1983) Nucleus tegmenti pedunculopontinus: efferent connections with special reference to the basal ganglia, studied by anterograde and retrograde transport of horseradish peroxidase. Neuroscience 10:725–765

Jacobsohn L (1909) Ueber die Kerne des menschlichen Hirnstamms. Verl Koenigl Akad Wissensch, Berlin

Jakob A (1923) Die extrapyramidalen Erkrankungen. Julius Springer, Berlin

Jia HG, Yamuy J, Sampogna S, Morales FR, Chase MH (2003) Colocalization of gamma aminobutyric acid and acetylcholine in neurons in the laterodorsal and pedunculopontine tegmental nuclei in the cat: a light and electron microscopic study. Brain Res 992:205–219

Johansson O, Hokfelt T, Pernow B, Jeffcoate SL, White N, Steinbusch HWM, Verhofstad AAJ, Emson PC, Spindel E (1981) Immunohistochemical support for three putative transmitters in one neuron: coexistence of 5-hydroxytryptamine, substance P- and thyrotropin releasing hormone-like immunoreactivity in medullary neurons projecting to the spinal cord. Neuroscience 6:1857–1881

Jones BE, Beaudet A (1987) Distribution of acetylcholine and catecholamine neurons in the brain stem: a choline acetyltransferase and tyrosine hydroxylase immunohistochemical study. J Comp Neurol 261:15–32

Kaiya T, Hoshino K, Norita M (2003) Postnatal development of cholinergic afferents from the pedunculopontine tegmental nucleus to the lateralis medialis-suprageniculate tegmental nucleus of the feline thalamus. Anat Embryol (Berl) 207:273–281

Kanazawa L, Marshall GR, Kelly JS (1976) Afferents to the rat substantia nigra studied with horseradish peroxidase, with special reference to fibers from the subthalamic nucleus. Brain Res 115:485–491

Kapadia SE, de Lanerolle NC (1984) Substance P neuronal organization in the median region of the interpeduncular nucleus of the cat: an electron microscopic analysis. Neuroscience 12:1229–1242

Kappers CUA, Huber CC, Crosby EC (1936) The comparative anatomy of the nervous system of vertebrates including man, vol II. Macmillan, New York

Karplus JP, Kreidl A (1909) Ueber die Bahn des Puppilar-reflexes (die reflektorische Puppilenstarre). Pflugers Arch Gesamte 49:115

Kearney JA, Albin RL (2000) Intra subthalamic nucleus metabotropic glutamate receptor activation: a behavioural, FOS immunocytochemical and [14C] 2-deoxyglucose autoradiographic study. Neuroscience 95:409–415

Keyser A (1972) The development of the diencephalon of the Chinese hamster. Thesis. Nijmegen
Kimura H, McGeer PL, Peng JH, McGeer EG (1981) The central cholinergic system studied by choline acetyltransferase immunohistochemistry in the cat. J Comp Neurol 200:151–201
Kita H, Kitai ST (1987) Efferent projections of the subthalamic nucleus in the rat: light and electron microscopic analysis with the PHA-L method. J Comp Neurol 260:435–452
Kita H, Chang HT, Kitai ST (1983) The morphology of intracellularly labeled rat subthalamic neurons: a light microscopic analysis. J Comp Neurol 215:245–257
Kitai H (2007) Globus pallidus external segment. Prog Brain Res 160:111–133
Kitai ST, Deniau JM (1981) Cortical inputs to the subthalamus: intracellular analysis. Brain Res 214:411–415
Kitai ST, Kita H (1987) Anatomy and physiology of the STN: a drive force of the basal ganglia. In: Carpenter MB, Jayaraman A (eds) The basal ganglia II. Structure and functions. Plenum Press, New York, pp 357–373
Kleiner-Fisman G, Herzog J, Fisman DN, Tamma F, Lyons KE et al (2006) Subthalamic nucleus deep brain stimulation: summary and meta-analysis of outcomes. Mov Disord 21:S290-S304
Knook HL (1965) The fibre connections of the forebrain. Thesis. Van Gorcum, Leiden, p 88
Kobayashi Y, Inoue Y, Isa T (2004) Pendunculo-pontine control of visually guided saccades. Prog Brain Res 143:439–445
Kodama S (1926) Ueber die sogenannten Basalganglien. Schweiz Arch Neurol Psychiat 19:152–177; 20:209–261
Kodama S (1927) Ueber die Entwicklung des striären Systems beim Menschen. Neurol Psych Abhand Heft 5:98
Kölliker A (1891) Zur feineren Anatomie des Centralnervensystems. Zeit Wiss Zool Bd 49–51:1–663
Kölliker A (1896) Handbuch der Gewebelehre des Menschen. Bd II, Leipzig
Kordower JH, Mufson EJ (2004) NGF receptor (p75)-immunoreactivity in the primate basal ganglia. J Comp Neurol 327:359–375
Kreiss DS, Anderson LA, Walters JR (1996) Apomorphine and dopamine D[1] receptor agonists increase the firing rates of subthalamic nucleus neurons. Neuroscience 72:863–876
Künzle H (1976) Thalamic projections from precentral motor cortex in Macaca fascicularis. Brain Res 105:253–267
Künzle H (1978) An autoradiographic analysis of the efferent connections from premotor and adjacent prefrontal regions (areas 6 and 9) in Macaca fascicularis. Brain Behav Evol 15:185–234
Künzle H (1998) Thalamic territories innervated by cerebellar nuclear afferents in the hedgehog tenrec, Echinops telfairi. J Comp Neurol 402:313–326
Künzle H, Akert K (1977) Efferent connections of cortical area 8 (frontal eye field) in Macaca fascicularis. A reinvestigation using the autoradiographic technique. J Comp Neurol 173:147–163
Kultas-Ilinsky K, Ilinsky IA (1990) Fine structure of the magnocellular subdivision of the ventral anterior thalamic nucleus (Vamc) of Macaca mulatta. II. Organization of nigrothalamic afferents as revealed with EM autoradiography. J Comp Neurol 294:479–489
Kultas-Ilinsky K, Ilinsky I, Warton S, Smith KR (1983) Fine structure of nigral and pallidal afferents in the thalamus: an EM autoradiography study in the cat. J Comp Neurol 216:390–405
Kultas-Ilinsky K, Leontiev V, Whiting PJ (1998) Expression of 10 GABA(A) receptor subunit messenger RNAs in the motor-related thalamic nuclei and basal ganglia of Macaca mulatta studied with in situ hybridization histochemistry. Neuroscience 85:179–204

Lakke EA (1997) The projections to the spinal cord of the rat during development; a time-table of descent. Adv Anat Embryol Cell Biol 135:143

Lakke EA, van der Veeken JG, Mentink MM, Marani E (1988) A SEM study on the development of the ventricular surface morphology in the diencephalon of the rat. Anat Embryol (Berl) 179:73–80

Lakke EA, Lazarov NE, Usunoff KG, Marani E (2000) A direct nigro-trigeminal projection in the rat: an anterograde tracer study using biotinylated dextran amine. Eur J Neurosci 12[Suppl 11]:135

Lankamp DJ (1967) The fibre composition of the pedunculus cerebri (crus cerebri) in Man. Thesis. Leiden University

Larsen M, Bjarkam CR, Ostergaard K, West MJ, Sorensen JC (2004) The anatomy of the porcine subthalamic nucleus evaluated with immunohistochemistry and design-based stereology. Anat Embryol (Berl) 208:239–247

Lavoie B, Parent A (1990) Immunohistochemical study of the serotoninergic innervation of the basal ganglia in the squirrel monkey. J Comp Neurol 299:1–16

Lavoie B, Parent A (1994a) Pedunculopontine nucleus in the squirrel monkey: distribution of cholinergic and monoaminergic neurons in the mesopontine tegmentum with evidence for the presence of glutamate in cholinergic neurons. J Comp Neurol 344:190–209

Lavoie B, Parent A (1994b) Pedunculopontine nucleus in the squirrel monkey: projections to the basal ganglia as revealed by anterograde tract-tracing methods. J Comp Neurol 344:210–231

Lavoie B, Smith Y, Parent A (1989) Dopaminergic innervation of the basal ganglia in the squirrel monkey as revealed by tyrosine hydroxylase immunohistochemistry. J Comp Neurol 289:36–52

Lee H, Choi BH (1992) Density and distribution of excitatory amino acid receptors in the developing human fetal brain: a quantitative autoradiographic study. Exp Neurol 118:284–290

Lee HJ, Rye DB, Hallanger AE, Levey AI, Wainer BH (1988) Cholinergic vs. noncholinergic efferents from the mesopontine tegmentum to the extrapyramidal motor system nuclei. J Comp Neurol 275:469–492

Leger L, Charnay Y, Hof PR, Bouras C, Cespuglio R (2001) Anatomical distribution of serotonin-containing neurons and axons in the central nervous system of the cat. J Comp Neurol 433:157–182

Levesque JC, Parent A (2005) GABAergic interneurons in human subthalamic nucleus. Mov Disord 20:574–584

Levey AI, Edmunds SM, Heilman CJ, Desmond TJ, Frey KA (1994) Localization of muscarinic m3 receptor protein and M3 receptor binding in rat brain. Neuroscience 63:207–221

Levin PM (1936) The efferent fibers of the frontal lobe of the monkey (Macaca mulatta). J Comp Neurol 63:369–419

Levy R, Hazrati LN, Herrero MT, Vila M, Hassano OK, Mouroux M, Ruberg M, Asensi H, Agid Y, Geger J, Obeso JA, Parent A, Hirsch EC (1997) Re-evaluation of the functional anatomy of the basal ganglia in normal and parkinsonian states. Neuroscience 76:335–343

Levy R, Ashby P, Hutchison WD, Lang AE, Lozano AM, Dostrovsky JO (2002) Dependence of subthalamic nucleus oscillations on movement and dopamine in Parkinson's disease. Brain 125:1196–1209

López-Giménez JF, Mengod F, Palacios JM, Vilaró MT (2001) Regional distribution and cellular localization of 5-HT2C receptor mRNA in monkey brain: comparison with [3H]mesulergine binding sites and choline acetyltransferase mRNA. Synapse 42:12–26

Luys J (1865) Recherches sur le système nerveux cérébro-spinal. J-B Ballière et Fils, Paris

Magill PJ, Bolam JP, Bevan MD (2001) Dopamine regulates the impact of the cerebral cortex on the subthalamic nucleus-globus pallidus network. Neuroscience 106:313–330
Mai JK, Assheuer J, Paxinos G (1997) Atlas of the human brain. Academic Press, San Diego
Mailleux P, Vanderhaeghen JJ (1992) Distribution of neuronal cannabinoid receptor in the adult rat brain: a comparative receptor binding radioautographic and in situ hybridization histochemistry. Neuroscience 48:655–688
Malinov GB, Usunoff KG, Romansky KV, Galabov GP (1984) Corticosubthalamic and corticonigral projection in the rat. Silver impregnation study. Acta Morphol Bulg 5:52–56
Mannen H (1960) "Noyau fermé" et "noyau ouvert". Arch Ital Biol 98:333–350
Marani E (1982) 5' Nucleotidase and acetylcholinesterase in the mammalian cerebellum. Thesis. Leiden University
Marani E, Schoen JHR (2005) A reappraisal of the ascending systems in man, with ephasis on the medial lemniscus. Adv Anat Embryol Cell Biol 179:76
Marani E, Usunoff KG, Feirabend HKP (2008) Lipofuscin and lipofuscinosis. In: Adelman G, Smith BH (eds) Encyclopedia of neuroscience. Elsevier, Amsterdam
Marburg O (1910) Mikroskopisch-topographischer Atlas des menslichen Zentralnervensystems. F Deuticke, Leipzig
Marburg O (1946) Das Striopallidum, seine Verbindungen und Bedeutung. Schweiz Arch Neurol Psychiatr 57:319–324
Marchand R (1987) Histogenesis of the subthalamic nucleus. Neuroscience 21:183–195
Marsden JF, Limousin-Dowsey P, Ashby P, Pollack P, Brown P (2001) Subthalamic nucleus, sensorimotor cortex and muscle interrelationships in Parkinson's disease. Brain 124:378–388
Martin GF, DeLorenzo G, Ho RM, Humbertson AO, Waltzer R (1985) Serotoninergic innervation of the forebrain in the North American oppossum. Brain Behav Evol 26:196–228
Martin JP (1927) Hemichorea resulting from a local lesion of the brain (the syndrome of the body of Luys). Brain 50:637–651
Martin-Cora FJ, Pazos A (2004) Autoradiographic distribution of 5-HT7 receptors in the human brain using [^{3}H]mesulergine: comparison to other mammalian species. Br J Pharmacol 141:92–104
Martinez A (1961) Fiber connections of the globus pallidus in man. J Comp Neurol 117:37–42
Maurice N, Deniau JM, Glowinski j, Thierry AM (1999) Relationships between the prefrontal cortex and the basal ganglia in rat: physiology of the cortico-nigral circuits. J Neurosci 19:4674–4681
McBride RL, Larsen KD (1980) Projections of the feline globus pallidus. Brain Res 189:3–14
McHaffie JG, Stanford TR, Stein BE, Coizet V, Redgrave P (2005) Subcortical loops through the basal ganglia. Trends Neurosci 28:401–407
Mehler WR (1981) The basal ganglia—circa 1982. A review and commentary. Appl Neurophysiol 44:261–290
Meibach RC, Katzman R (1979) Catecholaminergic innervation of the subthalamic nucleus: evidence for a rostral continuation of the A9 (substantia nigra) dopaminergic cell group. Brain Res 173:364–368
Meir A, Ginsburg S, Butkevich A, Kachalsky SG, Kaiserman I, Ahdut R, et al (1999) Ion channels in presynaptic nerve terminals and control of transmitter release. Physiol Rev 79:1019–1088
Mena-Segovia J, Bolam JP, Magill PJ (2004) Pedunculopontine nucleus and basal ganglia: distant relatives or part of the same family? Trends Neurosci 27:585–588
Mengod G, Nguyen H, Le H, Waeber C, Lubbert H, Palacios JM (1990) The distribution and cellular localization of the serotonin 1 C receptor m RNA in rodent brain examined by in

situ hybridisation histochemistry. Comparison with receptor binding distribution. Neuroscience 35:577–591

Mesulam MM, Mufson EJ, Wainer BH, Levey AI (1983) Central cholinergic pathways in the rat: an overview based on alternative nomenclature. Neuroscience 10:1185–1201

Mesulam MM, Mufson EJ, Levey AI, Wainer BH (1984) Atlas of cholinergic neurons in the forebrain and upper brainstem of the macaque based on monoclonal choline acetyltransferase immunohistochemistry and acetylcholinesterase histochemistry. Neuroscience 12:669–686

Mesulam MM, Geula C, Bothwell MA, Hersh LB (1989) Human reticular formation: cholinergic neurons of the pedunculopontine and laterodorsal tegmental nuclei and some cytochemical comparisons to forebrain cholinergic neurons. J Comp Neurol 283:611–633

Mettler FA (1945) Fiber connections of the corpus striatum of the monkey and baboon. J Comp Neurol 82:169–204

Mettler FA (1947) Extracortical connections of the primate frontal cerebral cortex. II. Corticofugal connections. J Comp Neurol 86:119–166

Mettler FA (1968) Anatomy of the basal ganglia. In: Vinken PJ, Bruyn GW (eds) Handbook of clinical Neurology, vol 6. Holland Publishing, Amsterdam, pp 1–55

Meyer M (1949) A study of efferent connexions of the frontal lobe in the human brain after leucotomy. Brain 72:264–296

Meyers R (1968) Ballismus. In: Vinken PJ, Bruyn GW (eds)Handbook of clinical neurology. North Holland Publishing, Amsterdam, pp 476–490

Miller WC, DeLong MR (1987) Altered tonic activity of neurons in the globus pallidus and subthalamic nucleus in the primate MPTP model of parkinsonism. In: Carpenter MB, Jayaraman A (eds) The basal ganglia II. Plenum Press, New York, pp 415–427

Milner TA (1991) Ultrastructural localization of tyrosine hydroxylase immunoreactivity in the rat diagonal band of Broca. J Neurosci Res 30:498–511

Misgeld U, Drew G, Yanovsky Y (2007) Presynaptic modulation of GABA release in the basal ganglia. Prog Brain Res 160:245–259

Mitchell IJ, Clarke CE, Boyce S, Robertson RG, Peggs D, Sambrook MA, Crossman AR (1989) Neural mechanisms underlying parkinsonian symptoms based upon regional uptake of 2-deoxyglucose in monkeys exposed to 1-methyl-4-phenyl-1,2,3,6-tetrahydropyridine. Neuroscience 32:213–226

Miwa H, Nishi K, Fuwa T, Mizuno Y (2000) Effects of blockade of metabotropic glutamate receptors in the subthalamic nucleus on haliperidol induced Parkinsonism in rats. Neurosci Lett 282:21–24

Miyata M (1986) Interconnections between the subthalamic nucleus and the cerebral cortex of the cat. Neurosci Res 4:1–11

Monteil A, Chemin J, Leuranguer V, Altier C, Mennessier G, Bourinet E, Lorry P, Nargeot J (2000) Specific properties of T-type calcium channels generated by the human α1 subunit. J Biol Chem 275:16530–16535

Moon Edley S, Graybiel AM (1983) The afferent and efferent connections of the feline nucleus tegmenti pedunculopontinus, pars compacta. J Comp Neurol 217:187–215

Morel A, Loup F, Magnin M, Jeanmonod D (2002) Neurochemical organization of the human basal ganglia: anatomofunctional territories defined by the distributions of calcium-binding proteins and SMI-32. J Comp Neurol 443:86–103

Morgan L0 (1927) The corpus striatum; a study of secondary degenerations following lesions in man and of symptoms and acute degenerations following experimental lesions in cat. Arch Neurol Psychiatr 18:495–549v

Morgan S, Huston JP (1990) The interhemispheric projection from the substantia nigra to the caudate-putamen as depicted by the anterograde transport of [^{3}H] leucine. Behav Brain Res 38:155–162

Mori S, Takino T, Yamada H, Sano Y (1985) Immunohistochemical demonstration of serotonin nerve fibers in the subthalamic nucleus of the rat, cat and monkey. Neurosci Lett 62:305–309

Mouroux M, Hassani OK, Feger J (1995) Electrophysiological study of the excitatory parafascicular projection to the subthalamic nucleus and evidence for ipsi- and contralateral controls. Neuroscience 67:399–407

Müller F, O'Rahilly R (1988a) The development of the human brain from a closed neural tube at stage 13. Anat Embryol (Berl) 177:203–224

Müller F, O'Rahilly R (1988b) The development of the human brain, including the longitudinal zoning in the diencephalons at stage 15. Anat Embryol (Berl) 179:55–71

Müller F, O'Rahilly R (1990) The human brain at stages 18–20, including the choroids plexuses and the amygdaloid and septal nuclei. Anat Embryol (Berl) 182:285–306

Nakamura S, Sutin J (1972) The pattern of termination of pallidal axons upon cells of the subthalamic nucleus. Exp Neurol 35:254–264

Nambu A, Llinas RR (1994) Electrophysiology of globus pallidus neurons in vitro. J Neurophysiol 72:1127–1139

Nambu A, Tokuno H, Takada M (1996) Functional significance of the cortico-subthalamo-pallidal 'hyperdirect' pathway. Neurosci Res 43:111–117

Nambu A, Tokuno H, Inase M, Takada M (1997) Corticosubthalamic input zones from forelimb representations of the dorsal and ventral divisions of the premotor cortex in the macaque monkey: comparison with the input zones from the primary cortex and supplementary motor area. Neurosci Lett 239:13–16

Nauta WJ (1958) Hippocampal projections and related neural pathways to the midbrain in the cat. Brain 81:319–341

Nauta WJ (1979) Projections of the pallidal complex: an autoradiographic study in the cat. Neuroscience 4:1853–1873

Nauta WJ, Cole M (1978) Efferent projections of the subthalamic nucleus: an autoradiographic study in monkey and cat. J Comp Neurol 180:1–16

Nauta WJ, Mehler WR (1966) Projections of the lentiform nucleus in the monkey. Brain Res 1:3–42

Ni Z, Bouali-Bennazzouz R, Gao D, Benabid AL, Bennazzouz A (2000) Changes in the firing pattern of globus pallidus neurons after the degeneration of nigrostriatal pathways are mediated by the subthalamic nucleus in the rat. Eur J Neurosci 12:4338–4344

Nieuwenhuys R, Voogd J, van Huijzen C (1988) The human central nervous system. A synopsis and atlas. Springer Verlag, Berlin Heidelberg New York

Nisbet AP, Forster OJ, Kingsburry A, Lees AJ, Marsden CD (1994) Nitric oxide synthase mRNA expression in human subthalamic nucleus, striatum and globus pallidus: implications for basal ganglia function. Brain Res Mol Brain Res 22:329–332

Nishimura Y, Takada M, Mizuno N (1997) Topographic distribution and collateral projections of the two major populations of nigrothalamic neurons. A retrograde labeling study in the rat. Neurosci Res 28:1–9

Nomura S, Mizuno N, Sugimoto T (1980) Direct projections from the pedunculopontine tegmental nucleus to the subthalamic nucleus in the cat. Brain Res 196:223–227

North RA (1993) Opioid actions on membrane ion channels. In: Herz A (ed) Opioids, vol 1. Springer-Verlag, Berlin Heidelberg New York, pp 773–798

Obeso JA, Olanow CW, Rodriguez-Oroz MC, Krack P, Kumar R, Lang AE (2001) Deep-brain stimulation of the subthalamic nucleus or the pars interna of the globus pallidus in Parkinson's disease. N Engl J Med 345:956–963

Ohishi H, Shigemoto R, Nakanishi S, Mizuno N (1993) Distribution of the mRNA for metabotropic glutamate receptor mGluR2 in the central nervous system of the rat. Neuroscience 53:1009–1018

Olszewski J, Baxter D (1954) Cytoarchitecture of the human brain stem. Karger, Basel, p 195
Orieux G, Francois C, Feger j, Yelnik J, Vila M, Ruberg M, et al (2000) Metabolic activity of excitatory parafascicular and pedunculopontine inputs to the subthalamic nucleus in a rat model of Parkinson's disease. Neuroscience 97:79–88
Otsuka T, Murakami F, Song WJ (2001) Excitatory postsynaptic potentials trigger a plateau potential in rat subthalamic neurons at hyperpolarized states. J Neurophysiol 86:1816–1825
Overton PG, O'Callaghan JF, Greenfield SA (1995) Possible intermixing of neurons from the subthalamic nucleus and substantia nigra pars compacta in the guinea-pig. Exp Brain Res 107:151–165
Pahapill PA, Lozano AM (2000) The pedunculopontine nucleus and Parkinson's disease. Brain 123:1767–1783
Papez JW (1942) A summary of fiber connections of the basal ganglia with each other and with other portions of the brain. Res Publ Assoc Res Nerv Ment Dis 21:21–68
Parent A (1996) Carpenters human neuroanatomy. William and Wilkens, Baltimore
Parent A (2002) Jules Bernard Luys and the subthalamic nucleus. Mov Disord 17:181–185
Parent A, Hazrati LN (1995a) Functional anatomy of the basal ganglia. I. The cortico-basal ganglia-thalamo-cortical loop. Brain Res Brain Res Rev 20:91–127
Parent A, Hazrati LN (1995b) Functional anatomy of the basal ganglia. II. The place of subthalamic nucleus and external pallidum in basal ganglia circuitry. Brain Res Brain Res Rev 20:128–154
Parent A, Smith Y (1987) Organization of efferent projections of the subthalamic nucleus in the squirrel monkey as revealed by retrograde labeling methods. Brain Res 436:296–310
Parent A, Mackey A, Smith Y, Boucher R (1983) The output organization of the substantia nigra in primate as revealed by a retrograde double labeling method. Brain Res Bull 10:529–537
Parent A, Cote PY, Lavoie B (1995) Chemical anatomy of primate basal ganglia. Prog Neurobiol 46:131–197
Parent A, Fortin M, Cote PY, Cichetti F (1996) Calcium-binding proteins in primate basal ganglia. Neurosci Res 25:309–334
Parent A, Parent M, Leroux-Hugon V (2002) Jules Bernard Luys: a singular figure of 19th century neurology. Can J Neurol Sci 29:282–288
Pasqualetti M, Ori M, Castagna M, Marazziti D, Cassano GB, Nardi I (1999) Distribution and cellular localization of the serotonin type 2C receptor messenger RNA in human brain. Neuroscience 92:601–611
Paxinos G, Huang XF (1995) Atlas of the human brainstem. Academic Press, San Diego
Pearson JC, Norris JR, Phelps CH (1985) Subclassification of neurons in the subthalamic nucleus of the lesser bushbaby (Galago senegalensis): a quantitative Golgi study using principal components analysis. J Comp Neurol 238:323–339
Perier C, Agid Y, Hirsch EC, Feger J (2000) Ipsilateral and contralateral subthalamic activity after unilateral dopaminergic lesion. NeuroReport 11:3275–3278
Perlmutter JS, Mink JW (2006) Deep brain stimulation. Annu Rev Neurosci 29:229–257
Petras JM (1965) Some fiber connections of the precentral and postcentral cortex with the basal ganglia, thalamus and subthalamus. Trans Am Neurol Assoc 90:274–275
Phelps PE, Brennan LA, Vaughn JE (1990) Generation patterns of immunocytochemically identified cholinergic neurons in rat brain stem. Brain Res 56:63–74
Philips T, Rees S, Augood S, Waldvogel H, Faull R, Svendsen C, Emson P (2000) Localization of metabotropic glutamate receptor type 2 in the human brain. Neuroscience 95:1139–1156
Pickel VM, Joh TH, Reis DJ, Leeman SE, Miller RJ (1979) Electronmicroscopic localization of substance P and enkephalin in axon terminals related to dendrites of catecholaminergic neurons. Brain Res 160:387–400

Pickel VM, Joh, TH, Chan J, Beaudet A (1984) Serotoninergic terminals: ultrastructure and synaptic interaction with catecholamine-containing neurons in the medial nuclei of the solitary tracts. J Comp Neurol 225:291–301

Pimlott SL, Pigott M, Greally E, Court JA, Jaros E, Perry RH, Perry EK, Wyper D (2004) Nicotinic acetylcholine receptor distribution in Alzheimer's disease, dementia with Lewy bodies, Parkinson's disease and vascular dementia: in vitro binding study using 5-[125I]-A-85380. Neuropsychopharmacology 29:108–116

Plenz D, Herrera-Marschitz M, Kitai ST (1998) Morphological organization of the globus pallidus-subthalamic nucleus system studied in organotypic cultures. J Comp Neurol 397:437–457

Pompeiano M, Palacios JM, Mengod G (1992) Distribution and cellular localization of mRNA coding for 5-HT-1A receptor in the rat brain: correlation with receptor binding. J Neurosci 12:440–453

Pompeiano M, Palacios JM, Mengod G (1994) Distribution of the serotonin 5-HT2 receptor family mRNAs: comparison between 5HT2A and 5-HT2C receptors. Brain Res Mol Brain Res 23:163–178

Poppi U (1927) Ueber die Fasersysteme der subst nigra. Arb Neurol Inst Wiener Univ B 29

Postuma RB, Lang AE (2003) Hemiballism: revisiting a classic disorder. Lancet Neurol 2:661–668

Prensa L, Parent A (2001) The nigrostriatal pathway in the rat: a single-axon study of the relationship between dorsal and ventral tier nigral neurons and the striosome/matrix striatal compartments. J Neurosci 21:7247–7260

Pritzel M, Sarter M, Morgan S, Huston JP (1983) Interhemispheric nigrostriatal projections in the rat: bifurcating nigral projections and loci of crossing in the diencephalon. Brain Res Bull 10:385–390

Quik M, Polonskaya Y, Gillespie A, Jakoweg M, Lloyd GK, Langston JW (2000) Localization of nicotinic receptor subunit mRNAs in monkey brain by in situ hybridization. J Comp Neurol 425:58–69

Rafols JA, Fox CA (1976) The neurons in the primate subthalamic nucleus: a Golgi and electron microscopy study. J Comp Neurol 168:75–111

Raiteri M (2006) Functional pharmacology in human brain. Pharmacol Rev 58:162–193

Rakonitz E (1933) Die Eigenerkrankung des Corpus Luysii. Der erste heredodegenerative Biballismus-Fall Z Gesamte. Neurol Psychiatr 144:255–266

Ramon y Cajal S (1955) Histologie Du Système Nerveux De L'Homme Et Des Vertébrés. Tome I. Généralités, Moelle, Ganglions Rachidiens, Bulbe Et Protubérance. Maloine, Paris. 1972. Segunda reimpression. Consejo Superior de Investigaciones Cientificas. Instituto Ramon y Cajal, Madrid

Ranson SW, Ranson M (1939) Pallidofugal fibers in the monkey. AMA Arch Neurol Psychiatry 42:1059–1067

Ranson SW, Ranson SW Jr (1941) The corpus striatum and thalamus of a partially decorticate monkey. AMA Arch Neurol Psychiatry 46:402–415

Ranson SW, Ranson Jr SW, Ranson M (1941) Fiber connections of the corpus striatum as seen in Marchi preparations. AMA Arch Neurol Psychiatry 46:230–249

Raynor K, Kong H, Mestek, A, Bye LS, Tian M, et al (1995) Characterization of the cloned human μ-opioid receptor. J Pharmacol Exp Ther 272:423–428

Reader TA, Sénécal J (2001) Topology of ionotropic glutamate receptors in brains of heterozygous and homozygous Weaver mutant mice. Synapse 42:213–233

Redgrave P, Marrow L, Dean P (1992) Topographical organization of the nigrotectal projection in the rat: evidence for segregated channels. Neuroscience 50:571–595

Richter E (1965) Die Entwicklung des Globus Pallidus und des Corpus Subthalamicum. Monogr. Gesamtgeb. Neurol. Psychiat, vol 108. Springer, Berlin Heidelberg New York

Richter EO, Hoque T, Halliday W, Lozano AM, Saint-Cyr J (2004) Determining the position and size of the subthalamic nucleus based on magnetic resonance imaging results in patients with advanced Parkinson disease. J Neurosurg 100:541–546

Riese W (1924) Ueber faseranatomische Verbindungen in "striaren System" der wasserlebende Sauger. Zeitschr Ges Neurol Psychiatr 90:591–598

Rinvik E (1968) The cortico-thalamic projection from the pericruciate and coronal gyri in the cat: an experimental study with silver impregnation methods. Brain Res 10:79–119

Rinvik E, Ottersen OP (1993) Terminals of subthalamonigral fibres are enriched with glutamate-like immunoreactivity: an electron microscopic, immunogold analysis in the cat. J Chem Neuroanat 6:19–30

Rinvik E, Grofová I, Hammond C, Feger J, Deniau JM (1979) A study of the afferent connections of the subthalamic nucleus in monkey and cat, using the HRP technique. In: Poirier LJ, Sourkes TL, Bedard PJ (eds) The extrapyramidal system and its disorders. Advances in neurology, vol 24. Raven Press, New York, pp 53–70

Robak A, Bogus-Nowakowska K, Szteyn S (2000) Types of neurons of the subthalamic nucleus and zona incerta in the guinea pig: Nissl and Golgi study. Folia Morphol (Praha) 59:85–90

Robledo P, Feger J (1990) Excitatory influence of rat subthalamic nucleus to substantia nigra pars reticulata and the pallidal complex: electrophysiological data. Brain Res 518:47–54

Rodriguez-Oroz MC, Rodriguez M, Guridi J, Mewes K, Chockkman V, Vitek J, DeLong MR, Obeso JA (2001) The subthalamic nucleus in Parkinson's disease: somatotopic organization and physiological characteristics. Brain 124:1777–1790

Romanowski CA, Mitchell JJ, Crossman AR (1985) The organization of the efferent projections of the zona incerta. J Anat 143:75–95

Romansky KV (1982) Structure and fiber connections of the subthalamic nucleus. PhD thesis, vol I:285, vol II:246. Medical Academy, Sofia

Romansky KV, Usunoff KG (1983) Electron microscopic identification of reticulo-subthalamic axon terminals in the cat. Neurosci Lett 42:113–117

Romansky KV, Usunoff KG (1985) The fine structure of the subthalamic nucleus in the cat. I Neuronal perikarya. J Hirnforsch 26:259–273

Romansky KV, Usunoff KG (1987) The fine structure of the subthalamic nucleus in the cat. II Synaptic organization. Comparisons with the synaptology and afferent connections of the pallidal complex and the substantia nigra. J Hirnforsch 28:407–433

Romansky KV, Usunoff KG, Ivanov DP (1978) Synaptic organization of the subthalamic nucleus in the cat. Etud Balk 31:1225–1228

Romansky KV, Usunoff KG, Ivanov DP, Galabov GP (1979) Corticosubthalamic projection in the cat: an electron microscopic study. Brain Res 163:319–322

Romansky KV, Usunoff KG, Ivanov DP (1980a) Vesicle containing dendrites in the subthalamic nucleus of the cat. J Hirnforsch 21:515–521

Romansky KV, Usunoff KG, Ivanov DP, Hassler R (1980b) Pallidosubthalamic projection in the cat. Electron microscopic study. Anat Embryol (Berl) 159:163–180

Rose JE (1942) The ontogenetic development of the rabbit's diencephalon. J Comp Neurol 77:61–130

Rouzaire-Dubois B, Scarnati E (1985) Bilateral corticosubthalamic nucleus projections: electrophysiological studies in rat with chronic cerebral lesions. Neuroscience 15:69–79

Rye DB, Saper CB, Lee HJ, Wainer BH (1987) Pedunculopontine tegmental nucleus of the rat: cytoarchitecture, cytochemistry, and some extrapyramidal connections of the mesopontine tegmentum. J Comp Neurol 259:483–528

Rye DB, Saper CB, Lee HJ, Wainer BH (1988) Medullary and spinal efferents of the pedunculopontine tegmental nucleus and adjacent mesopontine tegmentum in the rat. J Comp Neurol 269:315–341

Sadikot AF, Parent A, Francois C (1992) Efferent connections of the centromedian and parafascicular thalamic nuclei in the squirrel monkey: a PHA-L study of subcortical projections. J Comp Neurol 315:137–159

Salin P, Manrique C, Forni C, Kerkerian-Le Goff L (2002) High-frequency stimulation of the subthalamic nucleus selectively reverses dopamine denervation-induced cellular defects in the output structures of the basal ganglia in the rat. J Neurosci 22:5137–5148

Sano T (1910) Beitrage zur vergeleichenden Anatomie der Substantia nigra, des Corpus Luysii und der Zona incerta. Mschr Psychiat Neurol 27:110–127; 274–283; 381–389; 476–488; 28:26–34; 129–133; 269–278; 367–376

Saper CB, Loewy AD (1980) Efferent connections of the parabrachial nucleus in the rat. Brain Res 197:291–317

Sasaki T, Kennedy JL, Nobrega JN (1996) Autoradiographic mapping of μ opioid receptor changes in rat brain after long-term haloperidol treatment: relationship to the development of vacuous chewing movements. Psychopharmacology (Berl) 128:97–104

Sato F, Lavallee P, Levesque M, Parent A (2000a) Single-axon tracing study of neurons of the external segment of the globus pallidus in primate. J Comp Neurol 417:17–31

Sato F, Parent M, Levesque M, Parent A (2000b) Axonal branching pattern of neurons of the subthalamic nucleus in primates. J Comp Neurol 424:142–152

Satoh K, Armstrong DM, Fibiger HC (1983) A comparison of the distribution of central cholinergic neurons as demonstrated by acetylcholinesterase pharmacohistochemistry and choline acetyltransferase immunohistochemistry. Brain Res Bull 11:693–720

Schoen JHR (1964) Comparative aspects of the descending fiber systems in the spinal cord. Prog Brain Res 11:203–222

Schoen JHR (1969) The corticofugal projection on the brain stem and spinal cord in man. Psychiatr Neurol Neurochir 72:121–128

Schulz DW, Loring RH, Aizenman E, Zigmond RE (1991) Autoradiographic localization of putative nicotinic receptors in the rat brain using 125I-Neuronal Bungarotoxin. J Neurosci 11:287–297

Seifert G, Schilling K, Steinhauser C (2006) Astrocyte dysfunction in neurological disorders: a molecular perspective. Nat Rev Neurosci 7:194–206

Shen KZ, Johnson SW (2002) Presynaptic modulation of synaptic transmission by opioid receptor in rat subthalamic nucleus in vitro. J Physiol 541:219–230

Shik ML, Severin FV, Orlovsky GN (1966) Control of walking and running by means of electrical stimulation of the mid-brain. Biophysics 11:756–765

Shimamura K, Hartigan DJ, Martinez S, Puelles L, Rubinstein JLR (1995) Longitudinal organization of the anterior neural plate and neural tube. Development 121:3923–3933

Shink E, Bevan MD, Bolam JP, Smith Y (1996) The subthalamic nucleus and the external pallidum: two tightly interconnected structures that control the output of the basal ganglia in the monkey. Neuroscience 73:335–357

Shute CCD, Lewis PR (1967) The ascending cholinergic reticular system: neocortical, olfactory and subcortical projections. Brain 90:497–520

Skinner RD, Conrad N, Henderson V, Gilmore S, Garcia-Rill E (1989) Development of NADH diaphorase positive pecunculopontine neurons. Exp Neurol 104:15–21

Sloniewski P, Usunoff KG, Pilgrim Ch (1986) Retrograde transport of fluorescent tracers reveals extensive ipsi- and contralateral claustrocortical connections in the rat. J Comp Neurol 246:467–477

Smiley JF, Goldman-Rakic PS (1996) Serotonergic axons in monkey prefrontal cerebral cortex synapse predominantly on interneurons as demonstrated by serial section electron microscopy. J Comp Neurol 367:431–443

Smith Y, Kievel JZ (2000) Anatomy of the dopamine system in the basal ganglia. Trends Neurosci 23[10 Suppl]:S28–S33

Smith Y, Parent A (1984) Distribution of acetylcholinesterase-containing neurons in the basal forebrain and upper brainstem of the squirrel monkey (Saimiri sciureus). Brain Res Bull 12:95–124

Smith Y, Parent A (1988) Neurons of the subthalamic nucleus in primates display glutamate but not GABA immunoreactivity. Brain Res 453:353–356

Smith Y, Bolam JP, von Krosigk M (1990a) Topographical and synaptic organization of the GABA-containing pallidosubthalamic projection in the rat. Eur J Neurosci 2:500–511

Smith Y, Hazrati LN, Parent A (1990b) Efferent projections of the subthalamic nucleus in the squirrel monkey as studied by the PHA-L anterograde tracing method. J Comp Neurol 294:306–323

Smith Y, Wichmann T, DeLong MR (1994) Synaptic innervation of neurons in the internal pallidal segment by the subthalamic nucleus and the external pallidum in monkeys. J Comp Neurol 343:297–318

Smith Y, Bevan MD, Shink E, Bolam JP (1998) Microcircuitry of the direct and indirect pathways of the basal ganglia. Neuroscience 86:353–387

Smith Y, Kieval J, Couceyro PR, Kuhar MJ (1999) CART peptide-immunoreactive neurones in the nucleus accumbens in monkeys: ultrastructural analysis, colocalization studies, and synaptic interactions with dopaminergic afferents. J Comp Neurol 407:491–511

Smith Y, Charara M, Paquet M, Kieval JZ, Paré JF, Hanson JE, Hubert GW, Kuwijima M, Levey AI (2001) Ionotropic and metabotropic GABA and glutamate receptors in primate basal ganglia. J Chem Neuroanat 22:13–42

Somogyi P, Priestley JV, Cuello AC, Smith AD, Bolam JP (1982a) Synaptic connections of substance P-immunoreactive nerve terminals in the substantia nigra of the rat. A correlated light- and electron-microscopic study. Cell Tissue Res 223:469–486

Somogyi P, Priestley JV, Cuello AC, Smith AD, Takagi H (1982b) Synaptic connections of enkephalin-immunoreactive nerve terminals in the neostriatum: a correlated light and electron microscopic study. J Neurocytol 11:779–807

Song WJ, Baba Y, Otsuka T, Murakami F (2000) Characterization of Ca2+ channels in rat subthalamic nucleus neurons. J Neurophysiol 84:2630–2637

Spann BM, Grofová I (1992) Cholinergic and non-cholinergic neurons in the rat pedunculopontine tegmental nucleus. Anat Embryol (Berl) 186:215–227

Spatz H (1925) Ueber die Entwicklungsgeschichte der basalen Ganglien des menschlichen Grosshirns. Erg-H Anat Anz 60:54–103

Spatz H (1927) Physiologie und Pathologie der Stammganglien. Handb. Norm. Pathol. Physiol. Springer, Berlin Heidelberg New York

Steinbusch HW (1981) Distribution of serotonin-immunoreactivity in the central nervous system of the rat: cell bodies and terminals. Neuroscience 6:557–618

Steiner H, Weiler HT, Morgan S, Huston JP (1992) Time-dependent neuroplasticity in mesostriatal projections after unilateral removal of vibrissae in the adult rat: compartment-specific effects on horseradish peroxidase transport and cell size. Neuroscience 47:793–806

Steininger TL, Rye DB, Wainer BH (1992) Afferent projections to the cholinergic pedunculopontine tegmental nucleus adjacent midbrain extrapyramidal area in the albino rat. I Retrograde tracing studies. J Comp Neurol 321:515–543

Stenvers HW (1953) Clinical features of pyramidal and extrapyramidal disorders. Folia Psychiatr Neurol Neurochir Neerl 56:943–965

Sterman MB, Fairchild MD (1966) Modification of locomotor performance by reticular formation and basal forebrain stimulation in the cat: evidence for reciprocal systems. Brain Res 2:205–217

Strafella AP, Vanderwerf Y, Sadikot AF (2004) Transcranial magnetic stimulation of the human motor cortex influences the neuronal activity of subthalamic nucleus. Eur J Neurosci 20:2245–2249

Ströer WFH (1956) Studies on the diencephalon. I. The embryology of the diencephalon of the rat. J Comp Neurol 105:1–24
Strömberg I, Björklund H, Dahl D, Jonsson G, Sundström E, Olson L (1986) Astrocyte responses to dopaminergic denervations by 6-hydroxydopamaine and 1-methyl-4-phenyl-1,2,3,6-tetrahydropyridine as evidenced by glial fibrillary acidic protein immunohistochemistry. Brain Res Bull 17:225–236
Sugimoto T, Hattori T (1983) Confirmation of thalamo-subthalamic projections by electron microscopic autoradiography. Brain Res 267:335–339
Sugimoto T, Hattori T (1984) Organization and efferent projections of nucleus tegmenti pedunculopontinus pars compacta with special reference to its cholinergic aspects. Neuroscience 11:931–946
Sugimoto T, Hattori T, Mizuno N, Itoh K, Sato M (1983) Direct projection from the centromedian-parafascicular complex to the subthalamic nucleus in the cat and the rat. J Comp Neurol 214:209–216
Surmeier DJ, Bevan MD (2003) "The little engine that could" Voltage-dependent sodium channels and the subthalamic nucleus. Neuron 39:5–6
Svenningsson P, LeMoine C (2002) Dopamine D1/5 receptor stimulation induces c-fos expression in the subthalamic nucleus of local D5 receptors. Eur J Neurosci 15:133–142
Szabo J (1962) Topical distribution of the striatal efferents in the monkey. Exp Neurol 5:21–36
Szabo J (1967) The efferent projections of the putamen in monkeys. Exp Neurol 19:463–476
Szabo J (1970) Projections from the body of the caudate nucleus in the rhesus monkey. Exp Neurol 27:1–15
Szabo J (1972) The course and distribution of efferents from the tail of the caudate nucleus in the monkey. Exp Neurol 37:562–572
Takada M, Li ZK, Hattori T (1987) Long descending direct projection from the basal ganglia to the spinal cord: a revival of the extrapyramidal concept. Brain Res 436:129–135
Takada M, Nishihama MS, Nishihama CC, Hattori T (1988) Two separate neuronal populations of the rat subthalamic nucleus project to the basal ganglia and pedunculopontine tegmental region. Brain Res 442:72–80
Takakusaki K, Shiroyama T, Yamamoto T, Kitai ST (1996) Cholinergic and noncholinergic tegmental pedunculopontine projection neurons in rats revealed by intracellular labeling. J Comp Neurol 371:345–361
Talley EM, Cribbs LL, Lee JH, Daud A, Perez-Reyes E, Bayliss DA (1999) Differential distribution of three members of a gene family encoding low voltage activated (T-type) calcium channels. J Neurosci 19:1895–1911
Tanaka O, Sakagami H, Kondo H (1995) Localization of mRNAs of voltage dependent Ca^{2+}-channels: four subtypes of α1-and β-subunits in developing and mature rat brain. Mol Brain Res 30:1–16
Tande D, Feger J, Hirsch EC, Francois C (2006) Parafascicular nucleus projection to the extrastriatal basal ganglia in monkeys. Neuroreport 17:277–280
Temel Y, Blokland A, Steinbusch HWM, Visser-Vandewalle V (2005) The functional role of the subthalamic nucleus in cognitive and limbic circuits. Prog Neurobiol 76:393–413
Terada H, Nagai T, Okada S, Kimura H, Kitahama K (2001) Ontogenesis of neurons immunoreactive for nitric oxide synthase in rat forebrain and midbrain. Brain Res Dev Brain Res 128:121–137
Testa CM, Standaert DG, Young AB, Penney JB (1994) Metabotropic glutamate receptor mRNA expression in the basal ganglia of the rat. J Neurosci 14:3005–3018
Tintner R, Jankovic J (2002) Treatment options for Parkinson's disease. Curr Opin Neurol 15:467–476
Titica J, van Bogaert L (1946) Heredo-degenerative hemiballismus. Brain 69:251–263

Tokuno H, Nakamura Y, Kudo M, Kitao Y (1990) Laminar organization of the substantia nigra pars reticulata in the cat. Neuroscience 38:255–270

Touche M (1901) Deux cas d'hémichorée organique avec autopsie. Rev Neurol 9:1080

Tsou KT, Brown S, Udo-Pen MCS, Mackie K, Walker JM (1998) Immunohistochemical distribution of cannabinoid receptors in the rat central nervous system. Neuroscience 83:393–411

Turner TJ, Adams ME, Dunlap K (1993) Multiple Ca^{2+} channel types coexist to regulate synaptosomal neurotransmitter release. Proc Natl Acad Sci USA 90:9518–9522

Tuttle R, Braisted JE, Richards LJ, O'Leary DDM (1998) Retinal axon guidance by region-specific cues in diencephalons. Development 125:791–801

Usunoff KG (1984) Tegmentonigral projection in the cat. Electron microscopic observations. In: Hassler RG, Christ JF (eds) Parkinson-specific motor and mental disorders. Advances in neurology, vol 40. Raven Press, New York, pp 55–61

Usunoff KG (1990) Cytoarchitectural, ultrastructural and histochemical characteristics of the substantia nigra. Doctor Scientia Medicina Thesis, Sofia

Usunoff KG, Romansky KV (1983) Degenerating synaptic boutons in the substantia nigra and the subthalamic nucleus following destruction of the mesencephalic reticular formation. Verh Anat Ges 76:691–692

Usunoff KG, Hassler R, Romansky K, Usunova P, Wagner A (1976) The nigrostriatal projection in the cat. Part 1. Silver impregnation study. J Neurol Sci 28:265–288

Usunoff KG, Hassler R, Romansky KV, Wagner A, Christ JF (1982a) Electron microscopy of the subthalamic nucleus in the baboon. II. Experimental demonstration of pallidosubthalamic synapses. J Hirnforsch 23:613–625

Usunoff KG, Ivanov DP, Blagov ZA, Romansky KV, Malinov GB, Hinova-Palova DV, Paloff AU (1982b) Axonal degeneration following destruction of the mesencephalic reticular formation. III. Pathways arising in nucleus tegmenti pedunculopontinus and terminating in the monoaminergic neuronal groups of the midbrain, and in the basal ganglia. Med Biol Probl (Sofia) 10:27–42

Usunoff KG, Marani E, Schoen JHR (1997) The trigeminal system in man. Adv Anat Embryol Cell Biol 136:1–126

Usunoff KG, Itzev DE, Ovtscharoff WA, Marani E (2002) Neuromelanin in the human brain: a review and atlas of pigmented cells in the substantia nigra. Arch Physiol Biochem 110:257–369

Usunoff KG, Itzev DE, Lolov SR, Wree A (2003) Pedunculopontine tegmental nucleus. Part I. Cytoarchitecture, transmitters, development and connections. Biomedical Rev 14:95–120

van der Kooy D, Hattori T (1980) Single subthalamic nucleus neurons project to both the globus pallidus and substantia nigra in rat. J Comp Neurol 192:751–768

Ventura R, Harris KM (1999) Three-dimensional relationships between hippocampal synapses and astrocytes. J Neurosci 19:6897–6906

Verhaart WJ (1957) Fibre systems of the basal ganglia in systematic diseases. Proceed First Intern Congr Neurol Sciences, vol V:127–132

Verhaart WJC (1938a) A comparison between the corpus striatum and the red nucleus as subcortical centra of the cerebral motor system. Psych Neurol Bladen 5–6:1–60

Verhaart WJC (1938b) The rubrospinal system with monkeys and man. Psych Neurol Bladen 1938

Verhaart WJC (1950) Fibre analysis of the basal ganglion. J Comp Neurol 93:425–440

Verhaart WJC, Kennard MA (1940) Corticofugal degeneration following thermocoagulation of areas 4, 6, and 4 s in Macaca mulatta. J Anat 74:239–254

Vertes RP (1991) A PHA-L analysis of ascending projections of the dorsal raphe nucleus in the rat. J Comp Neurol 313:643–668

Vertes RP, Kocsis B (1994) Projections of the dorsal raphe nucleus to the brainstem: PHA-L analysis in the rat. J Comp Neurol 340:11–26
Vertes RP, Fortin WJ, Crane AM (1999) Projections of the median raphe nucleus in the rat. J Comp Neurol 407:555–582
Vilaró MT, Cortés R, Mengod G (2005) Serotonin 5-HT4 receptors and their mRNAs in rat and guinea pig brain: distribution and effects of neurotoxic lesions. J Comp Neurol 484:418–439
Villa M, Levy R, Herrero MT, Faucheux B, Obeso JA et al. (1996) Metabolic activity of the basal ganglia in parkinsonian syndromes in human and non-human primates: a cytochrome oxidase histochemistry study. Neuroscience 71:903–912
Villiger L (1946) Gehirn und Rueckenmark. B Schwabe and Co Verlag, Basel
Vincent SR (2000) The ascending reticular activating system—from aminergic neurons to nitric oxide. J Chem Neuroanat 18:23–30
Vincent SR, Satoh K, Armstrong DM, Fibiger HC (1983) NADPH diaphorase: a selective marker for the cholinergic neurons in the pontine reticular formation. Neurosci Lett 43:31–36
Vogt C, Vogt O (1920) Zur Lehre der Erkrankungen des striären Systems. J Psychol Neurol (Lpz) 25:627–846
Voigt MM, Laurie DJ, Seeburg PH, Bach A (1991) Molecular cloning and characterization of a rat brain cDNA encoding a 5-hydroxytryptamine 1B receptor. EMBO J 10:4017–4023
Von Monakow C (1895) Experimentelle und pathologisch-anatomische Untersuchungen ueber die Haubenregion, den Sehhuegel und die Regio subthalamica nebst Beitraege zur Kenntnis frueher-worbener Gross-und Kleinhirndefekte. Arch Psychiat Nervenkr 27:1–128; 386–478
Von Monakov C (1909) Der rote Kern, die Haube und die Regio hypothalamica bei einigen Säugetiere und beim Menschen. Arb hirnanatom Inst Zürich III: 53–267; IV:104–225
Voogd J, Feirabend HK (1981) Classic methods in neuroanatomy. In: Lahue R (ed) Methods in neurobiology. Plenum Press, New York, pp 301–364
Voogd J, Nieuwenhuys R, van Dongen PAM, ten Donkelaar HJ (1998) Mammals. In: Nieuwenhuys R, et al (eds) The central nervous system of vertebrates, vol 3. Springer, Berlin Heidelberg New York, pp 1635–2097
Waeber C, Dietl MM, Hoyer D, Palacios JM (1989) 5-HT1 receptors in the vertebrate brain. Naunyn Schmiedebergs Arch Pharmacol 340:486–494
Wang XS, Ong WY, Lee HK, Hugamir PL (2000) A light and electron microscopic study of glutamate receptors in the monkey subthalamic nucleus. J Neurocytol 29:743–754
Waselus M, Valentino RJ, Van Bockstaele EJ (2005) Ultrastructural evidence for a role of gamma-aminobutyric acid in mediating the effects of corticotropin-releasing factor on the rat dorsal raphe serotonin system. J Comp Neurol 482:155–165
Weiner DM, Levey AI, Brann MR (1990) Expression of muscarinic acetylcholine and dopamine receptor mRNAs in rat basal ganglia. Proc Natl Acad Sci USA 87:7050–7054
Whittier JR (1952) The graphic study of ballism and related hyperkinesia. J Neuropathol Exp Neurol 11:300
Whittier JR, Mettler FA (1949) Studies on subthalamus of rhesus monkey: hyperkinesia and other physiologic effects of subthalamic lesions, with special reference to the subthalamic nucleus of Luys. J Comp Neurol 90:319–372
Wiklund L, Descarries L, Mollgard K (1981) Serotoninergic axon terminals in the rat dorsal accessory olive: normal ultrastructure and light microscopic demonstration of regeneration after 5,6-dihydroxytryptamine lesioning. J Neurocytol 10:1009–1027
Williams MN, Faull RLM (1988) The nigrotectal projection and tectospinal neurons in the rat. A light and electron microscopic study demonstrating a monosynaptic nigral input to identified tectospinal neurons. Neuroscience 25:533–562

Wilson SA (1914) An experimental research into the anatomy and physiology of the corpus striatum. Brain 36:427–492

Wilson SA (1929) Die Pathogenese der unwillkuerlichen Bewegungen mit besonderer Beruecksichtigung der Pathologie und Pathogenese der Chorea. Dtsch Z Nervenheilk 108:4–38

Winkler C (1928) Handboek der Neurologie, Bouw van het zenuwstelsel I–V. Erven F. Bohn, Haarlem

Wisden W, Laurie DJ, Monyer H, Seeburg PH (1992) The distribution of 13 GABAA receptor subunit mRNAs in the rat brain. I. Telencephalon, diencephalon, mesencephalon. J Neurosci 12:1040–1062

Woelcke M (1942) Eine neur Methode der Markscheiden farbung. J Psychol Neurol 51:199–202

Woolf NJ, Butcher LL (1986) Cholinergic systems in the rat brain. III. Projections from the pontomesencephalic tegmentum to the thalamus, tectum, basal ganglia and basal forebrain. Brain Res Bull 16:603–637

Woolf NJ, Harrison JB, Buchwald JS (1990) Cholinergic neurons of the feline pontomesencephalon II. Ascending anatomical projections. Brain Res 520:55–72

Wright DE, Seroogy KB, Lundgren KH, Davis BM, Jennes L (1995) Comparative localization of serotonin 1A,1C and 2 receptor subtype mRNAs in rat brain. J Comp Neurol 351: 357–373

Xiang Z, Wang L, Kitai ST (2005) Modulation of spontaneous firing in rat subthalamic neurons by 5-HT receptor subtypes. J Neurophysiol 93:1145–1157

Yasui Y, Tsumori T, Ando A, Domoto T (1995) Demonstration of axon collateral projections from the substantia nigra pars reticulata to the superior colliculus and the parvicellular reticular formation in the rat. Brain Res 674:122–126

Yelnik J, Percheron G (1979) Subthalamic neurons in primates: a quantitative and comparative analysis. Neuroscience 4:1717–1743

Yokoyama CT, Westenbroek RE, Hell JW, Soong TW, Snutch TP, Catarall WA (1995) Biochemical properties and subcellular distribution of the neuronal Class E channel α1 subunit. J Neurosci 15:6419–6432

Yoshida M (1974) Functional aspects of, and the role of transmitter in the basal ganglia. Confin Neurol 36:282–291

Yu D, Gordon FJ (1994) A simple method to improve the reliability of iontophoretic administration of tracer substances. J Neurosci Methods 52:161–164

Zhang JH, Sato M, Tohyama M (1991) Region-specific expression of the mRNAs encoding beta subunits (beta 1, beta 2, and beta 3) of GABAA receptor in the rat brain. J Comp Neurol 303:637–657

Zhu J, Chen C, Xue JC, Kunapuli S, DeRiel JK, LiuChen LY (1995) Cloning of a human κ opioid receptor from the brain. Life Sci 56:201–207

Zhu Z, Bartol M, Shen K, Johnson SW (2002) Excitatory effects of dopamine on subthalamic nucleus neurons: in vitro study of rats pretreated with 6-hydroxydopamine and levodopa. Brain Res 945:31–40

Zubieta JK, Frey KA (1993) Autoradiographic mapping of M3 muscarinic receptors in rat brain. J Pharmacol Exp Ther 264:415–422

Index

Printing: Krips bv, Meppel, The Netherlands
Binding: Stürtz, Würzburg, Germany